Some Mathematical Questions in Biology. IX

Lectures on Mathematics in the Life Sciences

Volume 10

Some Mathematical Questions in Biology. IX

The American Mathematical Society

Providence, Rhode Island

1978

Proceedings of the Eleventh Symposium on
Mathematical Biology held at the Annual Meeting of the
American Association for the Advancement of Science
in Denver, Colorado, February, 1977.

edited by
Simon A. Levin

Library of Congress Catalog Card Number 77-25086
International Standard Book Number 0-8218-1160-6
International Standard Series Number 0075-8523
AMS 1970 Subject Classification: 92A05

The Symposium was sponsored by the
National Institutes of Health under Contract No. 263-76-C-0675

Printed in the United States of America
Copyright © 1978 by the American Mathematical Society

CONTENTS

FOREWORD

This volume contains lectures given at the Eleventh Sympo-
sium on Some Mathematical Questions in Biology, held in Denver,
Colorado on February 24-25, 1977, in conjunction with the Annual
Meeting of the American Association for the Advancement of
Science. The Symposium was co-sponsored by the American Mathe-
matical Society and by the Society for Industrial and Applied
Mathematics under the auspices of Section A, Mathematics, of the
AAAS.

The first two papers in the volume, by John Guckenheimer
and Giles Auchmuty, deal with mathematical issues which arise
from biological problems. Guckenheimer provides a critical
discussion of the uses and abuses of catastrophe theory (see also
J. Guckenheimer 1978, "The Catastrophe Controversy," The Mathematical Intelligencer $\underline{1}$:15-20 and the paper by E. C. Zeeman in
Volume 7 of this series), emphasizing both its success in geometrical optics and its failures elsewhere. He then goes on to
treat another class of nonlinear problems, those which give rise
to "chaotic" behavior. These, which have fascinated mathematicians and physicists for many years, have recently been of great
interest in population biology; and applications to the latter
provide a focus for Guckenheimer's paper.

Guckenheimer also discusses briefly bifurcation problems
associated with diffusion-reaction systems, and hence provides
a natural introduction to Giles Auchmuty's paper on the "Qualitative Effects of Diffusion in Chemical Systems." Diffusion-
reaction systems found wide application in biology: in popula-

tion biology, in developmental biology, in neurobiology, in epi-
demilogy, and in biochemistry. They have been discussed in
several articles in earlier volumes in this series. Auchmuty
summarizes recent work on such systems, including positivity of
solutions (invariance), existence, uniqueness, stability, bifurcation from steady state solutions and Hopf bifurcation of time
periodic solutions.

The third paper in the volume, by David Marr, confronts the
problem of the choice of representations for visual information.
Marr develops an overall framework for visual information processing, proceeding through three different representations
corresponding to different stages of analysis.

Problems in development provide the focus for the last two
papers in the volume, by Shymko and by Kauffman, Shymko and
Trabert. In the first of these, Shymko deals with the problem
of the regulation of growth, comparing two basic mechanisms,
control by (1) nutrient limitation and (2) by an internal "self-
regulatory" factor which acts at a critical point in the mitotic
cycle. He then suggests how these processes might be discriminated, and applies the method to experimental results with the
conclusion that the patterns apparently cannot be explained by
external factors alone. The suggested internal mechanism is by
means of a diffusible inhibitor secreted by the cells, and predictions based on this model are in good agreement with experiments.

The final paper, by Kauffman, Shymko, and Trabert is reprinted from Science, 199:259-270. A reaction-diffusion model in
which two diffusible biochemical components interact is proposed
for the control of sequential compartment formation during development of Drosophila melanogaster. The mathematical model
predicts a sequence of diffusive instabilities which correspond
to the geometries of compartment boundaries which subdivide the
early embryo and the imaginal discs. The chemical patterns

generate a binary combinatorial code and these are shown to
account for phenomena associated with transdetermination and
homeotic mutations.

It is a pleasure to acknowledge the support provided by the
National Science Foundation.

                                        Simon A. Levin

Comments on Catastrophe and Chaos

John Guckenheimer [1]

1. INTRODUCTION. The title of this lecture series is
"Some Mathematical Questions in Biology". In this lecture,
I take this title literally and raise a series of mathematical questions. Although many biologists might disagree, I hope that these are also problems which have some
relevance to the life sciences. The problems are ones of
"catastrophe" and "chaos" which have received considerable
attention in mathematical biology recently [28,40]. The
spectrum from catastrophe theory to chaos is a wide one,
and at times the organization of our remarks may appear
chaotic because it depends upon the underlying mathematical
ideas. I do not produce any new mathematical results here,
but only attempt to place existing work into perspective
and formulate a number of open questions. After an initial
broadly drawn discussion of catastrophe theory, we shall
concentrate on problems from population dynamics in discussing problems of chaos.

---

[1] Research partially supported by NSF Grant No. BMS74-21240

Two common threads weave themselves through our two topics. The first is a concern for "genericity" or "robustness". These two terms are used in mathematics and biology, respectively, to express similar concepts. The idea is that good models of a phenomenon should mimic it in a way that is not inappropriately sensitive to the details of the model. We have expressed this idea negatively to avoid the machinery necessary to state it precisely in a way suitable for all contexts. The best way of explaining it further is to provide an illustration of a model which is not robust.

The Lotka-Volterra equations for a predator-prey interaction in ecology are a pair of ordinary differential equations for the predator and prey populations [25]. The solutions of these equations predict that the predator and prey populations repeatedly return to their initial populations at later times, oscillating with an amplitude which varies with the initial populations. This behavior of solutions is not generic or robust for pairs of differential equations. The typical behavior of solutions is that they either tend to an equilibrium or a limit cycle. The limit cycle represents a population which tends to a regular periodic oscillation whose amplitude does not depend upon the initial populations. Unless there is a constraint - a law of ecology - which forces a predator-prey interaction to have the behavior predicted by the Lotka-Volterra equation, then we expect the prediction that all solutions return to their initial populations to be

wrong. The model itself is not robust within the class of
differential equations models appropriate for predator-prey
interactions. Considerations of this sort play a crucial
role in our discussions of both catastrophe theory and chaos.

The second thread which ties together the two parts of
the lecture is a concern for phenomena which are <u>intrinsically</u>
nonlinear. Much of applied mathematics is adapted to deal
with phenomena which can be described by linear models or
models which are approximately linear in some quantifiable
sense. The models we describe flaunt their nonlinearity by
making it an absolutely essential aspect of their behavior.
In some of the situations we shall be concerned with, linear
models fail the test of genericity or robustness. Their
qualitative properties are pathologic from our viewpoint.
This is primarily a mathematical consideration, but it seems
appropriate to biological phenomena which act in mysterious
(and highly nonlinear) ways. Whether these mathematically
motivated ideas will be translated into practical tools and
insights for biology is still very much an open question.
As we proceed, we focus on questions which appear directly
relevant to the problems of applying this seemingly abstruse
mathematics to biology.

I would like to thank George Oster for many stimulating
conversations over lunch at the Berkeley Faculty Club. His
critical comments on this manuscript have also been most
helpful.

II.   CATASTROPHE. "Catastrophe Theory" is the name which
has been given to the work of Thom [40] and subsequently
developed by others, particularly Zeeman [49]. We have reviewed the mathematical aspects of the theory some time ago
[11] and shall not dwell upon them here. Instead, we focus
upon the implications and applicability of the theory to
biology. The theory was initially created with applications
to developmental biology in mind, but the difficulties in
realizing applications throughout biology may be similar.
Our comments are quite general, largely because the chasm
between the mathematical theory and biological facts is
dismayingly wide. Moreover, the gulf appears to be closing
slowly, if at all. There are reasons for this state of
affairs, and we attempt to explain them. To do so, we must
recall very briefly the nature of catastrophe theory.

Catastrophe theory, interpreted broadly, is a viewpoint
from which particular problems are attacked. The subtitle
of Thom's book is "A General Theory of Models". Interpreted
narrowly, catastrophe theory consists of the theory of
singularities of smooth functions (<u>elementary</u> <u>catastrophe</u>
<u>theory</u>) and applications of this theory. Different exponents
of catastrophe theory have taken these very different viewpoints, so it is necessary to discuss both. Let us quickly
describe elementary catastrophe theory first.

The mathematics of the theory begins with a classification of the singularities of smooth functions. This clas-

sification collects together in one equivalence class those
functions (or germs of functions) which differ only by a
nonlinear change of coordinates in their domain. Precisely,
f, g: $\mathbb{R}^n \to \mathbb{R}$ have right-equivalent germs at 0 if there
is a map h: $\mathbb{R}^n \to \mathbb{R}^n$ defined on a neighborhood of $0 \in \mathbb{R}^n$
such that the Jacobian of h at 0 is nonsingular and
$f \circ h = g$ in a neighborhood of 0. The equivalence classes
for this classification represent functions which have
critical points at the origin with similar degeneracies.

Of particular interest is the unfolding of a singularity.
The unfolding is a family of functions (i.e., a function depending upon parameters) which is transverse to the equivalence class. The minimal number of parameters necessary to
construct such an unfolding is called the codimension of the
singularity. It is a measure of the degeneracy of the singularity. All of this is recounted in great detail elsewhere
[40], so we merely illustrate with an example. The function
f: $\mathbb{R} \to \mathbb{R}$ defined by $f(t) = \dfrac{t^4}{4}$ has a degenerate local
minimum at 0. The unfolding of f is given by
$f_{u,v}(t) = \dfrac{t^4}{4} + \dfrac{ut^2}{2} + vt$ where (u,v) are two parameters.
Any perturbation of f will be equivalent to one of the
functions $f_{u,v}$ in the sense described above. The unfolding of this function particular f is called the cusp
catastrophe.

The geometry associated with the cusp catastrophe has
been extensively used by Zeeman in applications. It is

illustrated in Figure 1. Let us paraphrase the argument
which is used to justify the ubiquitous cusp: very often
one does experiments in which one can simultaneously
manipulate two parameters. For some values of these parameters, the behavior of the experimental system may be discontinuous. Contained in the cusp catastrophe is the most
complicated generic way that such discontinuity can be present in a system depending upon two parameters. Therefore,
the cusp catastrophe (and the simpler fold) are a sufficient
tool kit for the modeler of discontinuous phenomena who re-
stricts himself to two experimental parameters. A good example of this sort of analysis is Zeeman's catastrophe
machine [48]. With this point of view, there has been a
tendency to see cusp catastrophes in a large number of
applications in widely diverse areas. Many of these applications are little more than vague analogies and have not
been pursued to the point of experimental verification.
There has been criticism of this work from different per-
spectives [8,38]. However, the fitting of cusp catastrophes
to discontinuous phenomena is not the point of view which
we would like to emphasize. The adequacy of these cusp
catastrophe models is not a mathematical question but rather
one of the usefulness of models based upon the study of
critical points of smooth functions. There is more to
catastrophe theory than the cusp, or even the hyperbolic
umbilic and butterfly.

Catastrophe theory, interpreted broadly, is one way of studying singular phenomena which occur in various situations. It proceeds by examining the generic singular phenomena (called catastrophes) which occur within the appropriate mathematical contexts. The meaning of the term "generic" depends upon the particular context of the model. Mathematically, one must construct a set of possible models and then examine the kinds of singular phenomena which occur in "almost all" models (as determined by a topological or measure theoretic structure for the set of models). In some instances, this leads to a reasonable classification of the singular phenomena or catastrophes of such a theory, while in other contexts this seems impossible. The singularity theory of smooth functions which leads to elementary catastrophe theory is the principal example of a "good" catastrophe theory. The philosophy underlying catastrophe theory is that a study of the geometry associated with the catastrophes in such a space of models will give clues as to whether the models themselves are good ones.

An immediate difficulty with this broad view of catastrophe theory is that there are few settings in which it has been possible to carry out the program of classifying the catastrophes. A decade ago the singularities of smooth maps were being classified in the work culminating with the thesis of Mather [26]. There was optimism that the bifurcations of vector fields could be described in much the

same way. Indeed, the specific connection between catastrophe theory and developmental biology in Thom's book depended
upon these hopes. Unfortunately, the optimism was not
justified: it became increasingly clear that describing
the qualitative changes of behavior in families of vector
fields was much messier than for families of functions. It
appears impossible that a reasonable multi-parameter bifurcation theory could be constructed. However, this does
not doom catastrophe theory to hopeless failure. There are
physical phenomena for which the applications of elementary
catastrophe theory are well understood (and useful) which do
not depend upon the bifurcation theory of vector fields.

For biology there remain substantial difficulties in
carrying through the vision of catastrophe theory in a
satisfying way. Part of the problem is that the catastrophes
which one expects to meet in a given context depend upon the
mathematical universe within which one works. It is not true
(as some expositions of catastrophe theory lead one to believe) that any discontinuity which one meets in the real
world must be one of the seven elementary catastrophes of
codimension at most four. To this assertion must be added
the qualification that it applies only when the appropriate
mathematical model is a generic family of smooth functions.
In many cases, this assumption is not true. Let us examine
a simple illustration of how the mathematical context within
which a model is constructed determines the predictions of

the model, even when we restrict ourselves to models based
upon the equivalence of smooth functions.

Consider the classification of smooth functions and
their unfoldings. If symmetries are imposed on the functions
being considered, new unfoldings and catastrophes arise [2].
For example, the unfolding x^6 in the class of even functions has a catastrophe set represented in the Figure 2.
Compare this unfolding with the 4-dimensional unfolding of
f as the "butterfly" catastrophe. Similarly, functions
which lie in a class of functions which are restricted to
be compositions of maps through an intermediate space can
have their own catastrophe theory [41]. This is illustrated
by the following example: the class of even functions on
the line coincides with the set of functions of the form
$f \circ g$ where $g(x) = x^2$. For this choice of g, compositions
of the form $f \cdot g$ have the same catastrophe theory as the
class of even functions. The example of x^6 illustrates
that the classification of catastrophes depends very strongly
upon the context within which one works. If the class of
functions which are considered "generic" for a particular
problem changes, then the "catastrophes" change as well.

This kind of example does not cause much difficulty in
those settings for which the class of functions to be considered is evident. There are many such problems. Zeeman's
catastrophe machine is a very good example. However, there
are also problems in the elastic stability of mechanical

structures for which it is not evident what the appropriate
class of functions should be [7]. This is particularly true
because we have a tendency to build structures which have a
high degree of symmetry. Nor is nature any different in
this regard. Biological organisms are full of structures
which are ordered in particular ways and show certain degrees of symmetry. In these problems, it is necessary to
have some means for deciding upon the appropriate class of
functions to use.

One possible answer to the problem of choosing a class
of functions in these situations is that symmetries tend to
be imperfect; so it will be rare that one should choose
anything other than the class of all functions. While this
may be literally true, it is an attitude which creates more
problems than it solves. In particular, if one considers a
family of functions which comes close to containing a degenerate singularity, then the bifurcation set of this family
will contain the unfoldings of several singularities close
to one another. The global geometry of the bifurcation set
now becomes crucial if one wants to understand all of the
catastrophes involved. This is illustrated by the following
example:

The function $f(x) = \frac{x^4}{4}$ has codimension 1 in the space
of functions invariant under reflection of the x-axis. The
space is the set of even functions, and the unfolding of f

is given by $f_u(x) = \frac{x^4}{4} + u\frac{x^2}{2}$. The bifurcation set of f
is pictured in Figure 3a. In the space of all functions f
is the cusp catastrophe. Its unfolding is pictured in
Figure 1. The unfolding of f in the space of even functions
is a degenerate family in the space of all functions. Nearby,
generic parameter families have unfoldings given in Figure 3b.
The generic perturbations of $f_u(x)$ in the space of all
functions have two branches which are close to one another.
This closeness is of practical importance if one takes the
point of view that degenerate singularities will not occur,
but only perturbations of them.

There are other ways in which restricted classes of
functions arise as models for physical phenomena. In con-
tinuum mechanics, for example, partial differential equations
provide detailed mathematical models. The bifurcations of
solutions of these equations may reveal considerably more
than an analysis which begins with a (Lagrangian) function
describing the state of the system. However, dealing directly with the partial differential equations is often a
difficult task. The situation is much the same in other
applications of catastrophe theory. If one obtains a function
describing a degenerate phenomenon from a reduction of a
more complicated mathematical description, there may well be
constraints on the functions one obtains from such reductions.
An illustrative example of this sort is the Holmes-Rand
analysis of periodic solutions of Duffing's equation [21].

Here a naive argument based upon elementary catastrophe
theory gave wrong answers at first. There is a moral here:
one should be constantly aware of the reductions already imposed upon a mathematical model when trying to give a catastrophe theoretic description of its discontinuous phenomena
with elementary catastrophe theory. The constraints inherent in a more detailed model can make a difference.

Next, let us examine those aspects of catastrophe theory
as they relate to biological morphogenesis. If one takes a
movie of a developing embryo, then certain apparent boundaries of tissues form as a result of differentiation. One
problem in applying catastrophe theory is to make a model of
the boundaries which allows one to describe their singularities. This may be a difficult modeling problem itself
if the boundaries are not sharp. Assuming that one can draw
a geometric picture of the tissue boundaries of a developing
organism, there are still other difficulties to overcome.
If the boundary singularities look like the bifurcation sets
of the elementary catastrophes, then we might consider the
problem solved: catastrophe theory says that these are the
expected singularities. However, if our singularities are
not the ones from catastrophe theory, then additional explanation is needed. This additional explanation must take
the form of a catastrophe theory which considers either a
restricted class of functions, or models which do not depend upon equivalence classes of functions. With physical

problems, the underlying differential equations often provide a convincing way of choosing another model, but with biological problems, determining the underlying equations is itself a task of enormous magnitude.

Despite these limitations, catastrophe theory still has something very useful to offer. The requirement that the catastrophe theories generated by models be consistent with geometric observations gives one a criterion for evaluating models! For example, the model which leads to elementary catastrophes may or may not be consistent with observations. If not, then we should discard models dependent upon generic families of smooth functions. This criterion from catastrophe theory is a means of eliminating models rather than one which helps in the construction of models or argues for the correctness of a specific model. The use of this criterion appears in work of Kleman-Toulouse [42] on phase transitions. They derive information about the symmetry properties of liquid crystals from the geometry of their order defects. A more complete analysis of this situation from the viewpoint of catastrophe theory would be of interest. What are the generic order defects of liquid crystals?

The point of view outlined in the last paragraph suggests a variety of mathematical questions whose solution might help in the understanding of morphogenesis. These problems involve a description of the bifurcation theories associated with systems of the sort which may be important in develop-

ment. For example, <u>what are the generic singularities which</u>
<u>occur in the solutions of "diffusion-reaction" systems of</u>
<u>equations</u> $u_t = f(u) + \Delta u$? (Here, $u: \mathbb{R}^n \times \mathbb{R} \to \mathbb{R}^n$,
$f: \mathbb{R}^n \to \mathbb{R}^n$, and Δ is the Laplacian.)

Beginning with Turing's classic paper [43], the role of
diffusion in pattern formation and morphogenesis has been a
subject of speculation and discussion. The discovery and
extensive study of the Belousov-Zhabotinsky reaction [47]
has further stimulated this discussion. If a diffusion-
reaction phenomenon is responsible for morphogenesis, then
the discontinuities of morphogenesis should correspond to
the generic singularities of a reaction-diffusion equation.

It may be that the singularities which arise from some
reaction-diffusion equations are described by elementary
catastrophe theory, but a glance at the Belousov-Zhabotinsky
reaction indicates that this will not always be so. There,
spontaneous patterns arise which have much more symmetry than
the elementary catastrophes. Another theory in which the
generic singularities may not be those predicted by elementary catastrophe theory is the study of shock waves in
systems of conservation laws. The study of conservation
laws is fundamental for theoretical understanding of such
biological techniques as chromatography [36]. <u>What is the</u>
<u>geometry of generic shock waves governed by quasi-linear</u>
<u>conservation laws</u>?

We pose one final question of this general kind: <u>how does the zero set of an eigenfunction of an elliptic partial differential operator on a bounded domain depend upon the boundary data</u>? This question is motivated by such biologically suggestive forms as "Chladni figures" on vibrating plates [22]. Is some form of potential field or standing vibration (described by an elliptic operator) involved in morphogenesis and regulation in development? Do the eigenfunctions of an elliptic operator show the same sort of dependence on boundary conditions as displayed by experiments in the morphogenesis of developing organisms? Evidence of this sort might be useful in solving some of the perplexing questions of biological pattern formation.

We have given a list of problems which <u>could</u> generate different catastrophe theories which <u>might</u> have relevance to the problem of morphogenesis. Without a better understanding of <u>both</u> the geometry of morphogenesis and the answers to a variety of mathematical questions, the relevance of catastrophe theory to problems of development will remain an open issue. Moreover, it is not clear that catastrophe theory can ever provide strong evidence for the correctness of any explanation of development. It may be limited to suggestive analogies between the mathematical theory and reality. Let us see how such suggestion can work for elementary catastrophe theory itself.

There is at least one area in which it is clear that

elementary catastrophe theory gives the "correct answer".
The propagation of singularities of solutions of a <u>linear</u>
hyperbolic partial differential equation is described by its
characteristic equation, a <u>nonlinear</u>, first order partial
differential equation. For example, the wave equation
$u_{tt} = \Delta_u$ is the basis of physical or wave optics, while
its characteristic equation $(u_t)^2 = \nabla u \cdot \nabla u$ (the eikonal
equation) is the basis for geometric optics. The characteristic equation can be thought of as describing the
propagation of "wavefronts" for the original equation. The
generic singularities of solutions of first order nonlinear
partial differential equations are described by elementary
catastrophe theory [1,14]. These are the singularities of
the wavefronts for the original problem. In optics they are
called <u>caustics</u>.

There are two sorts of mathematical questions raised
by the application of catastrophe theory to wave propagation problems. The first is its generality. For <u>linear</u>
hyperbolic equations, the description of singularities in
terms of catastrophe theory is satisfying. For <u>non-linear</u>
problems of wave propagation, this is not the case. Whitham
[46] has given a quite general formulation of non-linear
wave propagation problems for "slowly-varying waves". The
assumptions underlying these models are not likely to be met
at discontinuities. <u>Does elementary catastrophe theory play
a role in describing the singularities of nonlinear waves?</u>

One "classic problem of catastrophe theory is a geometric
description of breaking water waves. Thom states in his
book that breaking water waves have the geometry of a hyperbolic umbilic catastrophe, but no one has derived this
observation from a study of the Navier-Stokes equations of
fluid dynamics. Can elementary catastrophe theory be used to
motivate an exact theory of breaking waves? It seems unlikely.

The second problem raised by the applications of catastrophe theory to wave propagation looks at its success as
part of an approximation for a more detailed description of
the wave propagation. In optics (and the theory of linear,
hyperbolic partial differential equations generally), there
is a theory of uniform asymptotic approximations which deals
with this questions as wavelengths tend to zero. With biological situations, the wavelengths may not be small compared
to the scale of interest. What are the limitations of
asymptotic methods and elementary catastrophe theory in
solving linear hyperbolic partial differential equations?

Wave propagation may be an important process in morphogenesis. If so, the problems we have described are likely
to be important to an understanding of morphogenesis. However, if the singularities seen in morphogenesis do not
correspond to the elementary catastrophes, then it is unlikely that wave propagation plays a fundamental role. There
are other, more "global" questions which may also bear upon

morphogenesis and regulation in biological organisms. Catastrophe theory, as it applies to biology, begins with the
assumption that the "geometry" of animals is qualitatively
stable. It then attempts to examine the consequences of
this assumption for models of morphogenesis. In my opinion,
a fundamental difficulty with the theory as represented by
Thom's "static model" is that it does not sufficiently take
into account the interaction of processes at different points
of an organism. Without further assumptions, this deficiency
precludes making predictions about anything other than the
local shape of discontinuities. If one views wave propagation of some sort as a mechanism for introducing catastrophes, then there is the possibility of studying the regulation of morphogenesis by examining the influence of
existing boundaries on the wave propagation process. This
viewpoint suggests mathematical questions about the necessary
complexity of catastrophe sets, about which little is known.
One small example is an analysis of Zeeman's catastrophe
machine [48].

Before passing to questions of chaos, let us end our
comments about the role of catastrophe theory in biology
with a summary of our assessment of the success of the theory
ten years after its creation. There clearly have been un-
realized hopes. The philosophical ideas of genericity and
structural stability to which catastrophe theory has
brought great attention have changed the way in which cer-

tain problems have been approached. On this level, catastrophe theory may indeed be a scientific revolution of the
sort claimed by its most vigorous proponents. On the other
hand, attempts to explain morphogenesis in terms of the
elementary catastrophes have not been more convincing than
the gradient theories they were intended to supplant. The
"mechanism" of morphogenesis and spatial regulation in
biological organisms remains an unsolved mystery. Catastrophe theory has not significantly enhanced our understanding of these processes. It may still do so, but theories
based upon the cusp catastrophe appearing in two parameter
families of functions are not likely to provide the missing
clue to the mystery. Much more likely in my opinion is that
the key will be in studies of the "catastrophe theory"
associated to biologically meaningful systems of equations
(such as diffusion-reaction equations). Still more likely
perhaps is that the answer to the problem of pattern formation will come from entirely different directions.
Whether or not "waves" are an important part of the process
of morphogenesis seems to be a more reasonable biological
question at this time than the question of whether the
elementary catastrophes are models for the singularities of
morphogenesis.

Speculating from an entirely different viewpoint, one
possibility is that a study of simple algorithms may prove
to be useful in understanding morphogenesis. Consider the

problem of how an ant colony builds its nest. We observe
the behavioral repertoire of individual ants and try to find
a procedure for constructing the nest from these basic units
of behavior. Specifying an algorithm for determining how
individual ants determine which acts they perform at any
given time seems like a reasonable approach to this problem.
Teaching a computer to build an ant nest may be a useful
task in understanding morphogenesis. Nonetheless, this
approach also calls for caution since purely combinatorial
approaches to morphogenesis have been no more successful than
those of catastrophe theory.

III.   CHAOS. Let us turn now to a different class of in-
trinsically nonlinear problems which arise when one examines
dynamic processes in biology. We shall focus our attention
on population dynamics since it is the context within which
these problems have been raised. While the biological problems involved are very different from those considered
above, the mathematical ideas of genericity and structural
stability again provide a good philosophical attitude from
which the problems can be approached. May [28] has recently
reviewed the biological literature surrounding the questions
of "chaos" which we discuss critically here.

To begin our discussion, we examine a particular difference equation which can be used to illustrate the dynamic
phenomena which occur in some population models. Consider
a single population with discrete generations which obeys

a simple law of density dependent population growth. We
assume that this law can be expressed by an equation of the
form $N_{t+1} = f(N_t)$ where N_t is the population at time t
and f: $\mathbb{R} \to \mathbb{R}$. The choice of the function f reflects
assumptions about the ecology of the species being investigated. Many functions have been used in the biological
literature: May and Oster [29] review models of this sort.
One particular choice of f which has received substantial
attention is $f(x) = 4\mu x (1-x)$; $x,\mu \in [0,1]$. We shall not
attempt to relate this or any other function to a particular
biological situation, rather we consider it for purposes of
illustrating the sorts of dynamic phenomena which occur in
more realistic population models.

The problem of population dynamics is to describe the
sequence N_t if N_0 is known. This is the mathematical
problem of describing the sequences $\{f^i(x)\}_{i \geq 0}$ for each
$x \in I$. Here the superscript denotes repeated composition of
f with itself, and the set $\{f^i(x)\}$ is called the <u>orbit</u>
of x.

It has long been known that the orbits of f can be
very complicated. Of special interest are orbits with the
property that $f^k(x) = x$ for some i > 0. These orbits
are called <u>periodic</u>. The smallest i > 0 such that
$f^i(x) = x$ is the <u>period</u> of x. For μ close to 1,
$f(x) = 4\mu x (1-x)$ has an infinite number of periodic orbits.
There are also orbits which do not tend to a periodic orbit.

When $\mu = 1$, there is more to be said. The map
$f(x) = 4 x (1-x)$ has a dense set of periodic orbits. Most
orbits do not tend to any sort of limit but are sequences
which are dense in the interval. If the two halves of the
interval $[0,\frac{1}{2}]$ and $[\frac{1}{2},1]$ are thought of as "states" of
the population (say, the increasing state and the decreasing
state) then knowledge of the current state of the population
provides no information about its subsequent states. In this
sense, the population appears to behave in a random way.
This typifies the behavior associated with the term "chaos"
in the recent biological literature.

Another aspect of interest for the functions
$f_\mu(x) = 4\mu x (1-x)$ is the dependence upon μ. Changes in
the qualitative features of the dynamics of a map are called
bifurcations. One type of bifurcation of the map $f_\mu(x)$
involves a change in the number of periodic orbits of a
given period. These bifurcations have been studied also.
There is a well determined (but complicated) pattern according to which new periodic orbits appear as μ increases
[17]. For each value of μ, there is at most one periodic
orbit which is stable in the sense that orbits close to the
periodic orbit tend to it. For some values of the parameter
μ, the period of this stable periodic orbit is very sensitive
to small changes in μ. This illustrates another phenomenon
common in population models: qualitative features of the
dynamics of a model may be very sensitive to small changes

in parameters entering into the model.

This simple one dimensional model is not typical of the kind of behavior one expects to see from "chaotic" models in some ways. For a large set of the parameter values, most initial conditions do tend asymptotically to a stable periodic orbit. In more elaborate models, complicated limiting behavior can be the rule for almost all choices of initial values in some region of the state space. Ruelle and Takens have called these limit sets "strange attractors" in applying similar ideas to the study of turbulent fluid motion. There is some confusion as to what the definition of a strange attractor should be. For our purposes, the important properties of a strange attractor are (1) that it be "robust" in some suitable sense, and (2) that nearby initial conditions usually have orbits which diverge from one another.

There is a class of strange attractors which have been extensively studied. These are the "Axiom A" strange attractors introduced by Smale. In order to focus the discussion, it is worthwhile to introduce one of these attractors and then to discuss some of their properties. Only then will we be able to return to the biological issues of chaos that we are to consider. The questions we raise make little sense in the absence of a clear picture of one of these models.

The strange attractor we define was first described by Smale [37]. It lies inside a solid torus T in 3 space

and is constructed as follows. Define a map f: T → T which
stretches the torus along its center line, contracts it in
the transverse direction, and then places the long, thin torus back inside T by wrapping it around twice. See Figure
4a. The map f defined geometrically in this way has a
strange attractor Λ lying inside T. This attractor consists of the points in the intersection $\cap_{i \geq 0} f^i(T)$. To see
what the attractor looks like, slice the attractor by a two
dimensional "transverse" disk D. Sitting inside this disk
are its insections with the successive images of T under
iterates of f. See Figure 4b. The set $\Lambda \cap D$ is a "Cantor
set". Mathematically, it is a closed, zero dimensional, un-
countable set. The set Λ will be a 1 dimensional set
(called a solenoid) whose intersection with any transverse
disk in T is a Cantor set. Any point T has an orbit which
tends to the set Λ, but its limit behavior in Λ is unpre-
dictable in some aspects. The combination of partial pre-
dictability and partial unpredictability raises a number of
questions to which we shall soon return.

      The attractor Λ that we have just introduced sat-
isfies Smale's Axiom A. Given a map f: M → M, Axiom A states
that, along a strange attractor, there is a "good" splitting
of M into directions along which f is contracting and those
along which f is expanding. In the example above, f is
expanding in a direction along the center circle of the torus,
and f contracts transverse disks. The implications of Axiom
A are of two sorts. First, Axiom A attractors are

structurally stable in that perturbations of the dynamical
system will yield attractors which have the same geometric
properties as the original one. Second, Axiom A attractors
which are larger than a single periodic orbit have very
strong ergodicity properties. There is an elegant "statistical mechanics" for Axiom A attractors which describes
the statistical properties of orbits approaching the attractor. Thus Axiom A attractors are the archetype of
strange attractors. They have been extensively studied
from a mathematical point of view and display in the strong-
est possible way the "chaotic" behavior which can be expected from population models [4].

There remain many mathematical questions about the place
of strange attractors in models. These questions can be
addressed by considering maps of the plane into itself. A
population model of this sort which has been studied from
this point of view is the map $f: \mathbb{R}^2 \to \mathbb{R}^2$ defined by
$f(x_1,x_2) = ((b_1 x_1 + b_2 x_2)e^{-(x_1+x_2)}, x_1)$ [18]. One can ask
whether this map f has a strange attractor for a large
set of values for (b_1,b_2). Numerical simulations suggest
that the answer is yes. For a large set of values for
(b_1,b_2), most initial conditions have orbits which do not
tend to periodic orbits of small period in many thousands
of iterations. On the other hand, there is theoretical
evidence [32] that maps of this sort should typically have
stable periodic orbits of very long period. These will be

distributed in the plane in much the same way as if they were part of a strange attractor. The regions of attraction of points in these stable periodic orbits are likely to be very small.

A more complete mathematical analysis of the bifurcation behavior of maps of the plane such as the one above would help sort out the issues of when models can be expected to have strange attractors (according to some precise definition of a strange attractor). Even for the one dimensional map $f_\mu(x) = 4\mu x(1-x)$, the following question is unresolved: is there a set of positive measure in the parameter interval such that the corresponding f_μ have sets of positive measure consisting of orbits which do not tend asymptotically to a stable periodic orbit ? A two dimensional example which has no critical points has been studied numerically by Henon: $f(x_1,x_2) = (x_2, bx_1 + 4\mu x_2(1-x_2))$. Questions of the existence of strange attractors and the bifurcation behavior of maps in the plane arise naturally in the study of forced, non-linear oscillations [20]. Indeed, the study of chaotic behavior in non-linear oscillations provided inspiration for the modern development of dynamical systems [5,24]. While there are examples of maps of the plane into itself which do have Axiom A strange attractors, [35] the question of whether a forced non-linear oscillation or a population model could give rise to such a map remains open.

These questions have more of a theoretical rather than

a practical interest for biology. Whether points lie in
weakly attracting periodic orbits of very long period or in
an actual strange attractor is unimportant for population
models of the sort described above. The question of time
scale here is important. With models that predict varying
populations, fluctuations in climate and environment on the
one hand and genetic changes in the population on the other
make precise predictions about the size of a population
after hundreds or thousands of generations a hazardous
business. The relevant question for population biology
is whether models which appear to have strange attractors
on the time scales of interest reasonably reflect the reality
of population dynamics. Some attempts to fit models to
data [19] have led to the conclusion that chaotic models do
not work well for most systems. I disagree.

There are two different "real worlds" a scientist can
look toward in trying to resolve this question. The first
is the world of laboratory ecosystems; the second is the
"field". Field studies seldom contain data which are extensive enough to compare with a detailed model. Moreover,
fluctuations of climate and environment make it difficult,
if not impossible, to determine what the essential features
of a model should be. In the laboratory, one can control
the environment and sampling errors at the sacrifice of
studying the system in its natural biological setting. It is
certainly easier to test chaotic models against the population

dynamics of laboratory ecosystems as compared to natural
ecosystems. We want to discuss this issue here, indicating
a number of inherent problems in resolving the issue in a
definitive way.

The simplest sort of experiment which could be performed
is to grow a single species in a constant environment. Experiments of this sort were performed by A.J. Nicholson in
Australia during the 1950's. His experiments showed that
blowfly populations grown in constant conditions could
fluctuate dramatically. Figure 5 shows the population as
a function of time in a typical experiment of Nicholson.
Nicholson's data shows rather irregular fluctuations. One
can ask whether the blowflies are governed by equations
whose solutions are periodic [27], or whether a model with
deterministic chaos is more appropriate. Are the differences
between Nicholson's data and a periodic function due to
random influences, or is there more information in the data
which is implicit in an underlying deterministic model?
This sort of question seems to me to be fundamental in
population biology. Its answer will determine in large part
the amount of data necessary to make reasonable comparisons
with models. It is easier to estimate the period and
amplitude of a periodic oscillation than to describe the
feature of more irregular fluctuations.

The answer to this last question is likely to be that
chaotic models are appropriate sometimes. Indeed, chaotic

models seem to be appropriate for Nicholson's experiment.
The minimal ingredients that a detailed model for these experiments should include are the age-structure of the population and the "density-dependent" effects of competition
on the vital parameters of the population [34]. Models of
this sort with realistic (measured) parameters do have
chaotic behavior on the time scales of interest. Ongoing
efforts with G. Oster, D. Auslander, C. Wu, A. Ipaktchi, and
the author indicate that age structured models with chaotic
behavior give a much better approximation to blowfly population data than a model with regularly periodic behavior
could. Realizing that the issue of the reality of deterministic chaotic behavior in population biology is still
unsettled [19], we shall assume nevertheless that the phenomenon is important and discuss its implications with this
presumption in mind.

Perhaps the most serious set of questions revolves
around the issue of how a comparison is to be made between
experimental data and a chaotic model. With many mathematical models (e.g. fitting a straight line to a set of data),
there are one or a few parameters to choose in selecting the
model. The statistical criteria for making this choice are
a matter of well accepted convention. This is certainly not
the case with a chaotic model. The difficulty is reflected
in the fact that nearby initial conditions will have orbits
which diverge from one another in time. This results in a

situation in which the predictions of the model are very
sensitive to initial conditions. Moreover, small random
fluctuations introduce differences which grow with time.
Assessing whether a particular set of data could reasonably
have come from a certain model may be a difficult and time
consuming enterprise even if it is theoretically possible.
A second difficulty is reflected in the fact that chaotic
models may have many parameters, and the dynamics of individual orbits may be as sensitive to parameters as initial
conditions. This is evident in the age structured models
which have been used to simulate blowfly experiments. Thus
we ask the question: <u>Are there mathematical procedures which
allow one to (efficiently) locate the parameters and initial
conditions which give rise to an orbit generated by a chaotic
model?</u>

Even with procedures allowing one to choose parameters
and initial conditions, one would like to have statistical
criteria for comparing experimental data with alternate
models. We have asked already whether a model with regular
oscillations and random fluctuations is a better model for
Nicholson's experiments than one which is chaotic in a
deterministic way. This question makes little sense in the
absence of statistical criteria for making comparisons.
These questions are fundamental in applying mathematical
models in the study of fluctuating populations. Without
guidelines of some sort, one is left in the position of

making subjective judgements about the adequacy of models to
reproduce the data. There is no basis for preferring one
model to another. If data appears to have features of both
regularity and irregularity, then a model which "explains"
all of the irregularity on the basis of random influences
may miss an essential part of the biology. On the other
hand, such a model may be correct. In principle, it is probably impossible to give a definitive means for picking the
most appropriate model. Nonetheless, practical guidelines
based on extensive experience with population models of
various kinds might be obtainable.

Let us carry this discussion still further. One expects
that in controlled population experiments the data will be
well approximated by a deterministic model. Differences between the predictions of the "best" deterministic model and
the data should be due to random effects. <u>When the appropriate deterministic models are chaotic, how does one separate
the deterministic and stochastic components of the data?</u>
There are theoretical reasons for thinking that this question
is irresolvable in principle. Axiom A attractors have the
property that the addition of small random perturbations
has little effect upon their dynamics. In some ways, their
chaotic behavior is already as random as flipping a coin
repeatedly. It is hard to imagine that their dynamics could
be more random by the addition of "noise". This creates a
real dilemma, because one does want to discover the under-

JOHN GUCKENHEIMER

lying laws of population dynamics as represented by the best deterministic model. In the physical sciences, these problems are usually dealt with by doing many trials of one experiment. With population biology, it does not seem possible to do this in the field and doing it in the labora- tory is a formidable task. Biologists may have to accept a theoretical inability to resolve data into its underlying deterministic and stochastic effects just as physicists were forced to accept an indeterminacy principle.

Issues similar to the ones being considered here arise in the study of mathematical models for the turbulent flow of fluids. There, different models give different predic- tions about certain statistical properties of the velocity of a turbulent fluid at a particular point. These predictions concern the "spectral" properties of the velocity as deter- mined by Fourier analysis [10]. One can ask such questions of population models. What is the role of time series anal- ysis in population biology? Is Fourier analysis a useful way to discriminate the predictions of various models and make comparisons with data? It is certainly plausible that spectral analysis will provide a means of quantitatively assessing the reality of "cycles" in population biology. A major difficulty that arises in this endeavor is the avail- ability of data of sufficient duration and detail to make time series analysis meaningful.

The problems entailed in collecting detailed, long term

data brings us to our final topic. No essay on grand schemes
of biology would be complete without a discussion of evolution. The relationship between chaotic population dynamics
and evolutionary theory is a fascinating one which begs for
more study than we can give it here. Nicholson found that
evolution played a very important role in his experiments
with blowflies and wrote a very provocative paper on the
subject [31]. We shall discuss some of these same issues
from the perspective of chaotic population models, asking
what their impact upon the theories of population genetics
should be.

Classical theories of population genetics can be ca-
ricatured as "urn" models [9]. Populations represent urns
which are filled with balls of different colors representing organisms having different alleles for one or more genes.
The game of "life" is generally played in terms of discrete
generations, with certain rules as to how current frequencies of balls of different colors affect the distribution of
colors in the next generation. These rules are the laws of
genetics. Theories of this classical sort have worked well
in laboratory or breeding experiments in which the population
dynamics were artificially controlled. However, when con-
fronted with data from natural populations, the classical
theories have not worked well. For example, they seriously
underestimated the extent of genetic polymorphism in natural
populations [20]. Some attempts to deal with these issues

have looked toward frequency and density dependent selection as the principal key to its resolution [30]. It is my contention that the interaction of the population dynamics and the genetic structure of a population do lead to a significantly altered perspective on evolutionary theory from the one put forward in the classical theories of population genetics.

The presence of chaos in population dynamics results in a situation in which the conditions faced by organisms of a species change radically in time. If the vital parameters for the species are subject to genetic variation, then it may happen that different alleles are advantageous at different times when the population faces different conditions. The result of the varying conditions created by the dynamic fluctuations may be that no allele becomes extinct because each will face favorable conditions before it is eliminated. Moreover, the gene frequencies and the population dynamics may be interrelated so that both fluctuate and influence the other. There need be no genetic equilibrium just as there is no population equilibrium.

The concept of optimization plays a large role in much of the thinking about evolution. Indeed the phrase "survival of the fittest" is often transformed into the idea that natural selection works to maximize fitness. Fitness is given a quantitative meaning, typically as the number of viable offspring. It is easy to see how selection works to maximize fitness of this sort in an urn model or in an experiment in which mortality and fecundity depend only on genetic structure. These arguments can be summarized by the principle that one organism that does something better than another will be selected, all other things being equal. The qualification is crucial because all other things are seldom equal.

A series of experiments conducted by C. Huffaker and his colleagues illustrates how this scheme might work in a simple situation [44]. Imagine an insect host and parasite living in a constant environment. The parasite seeks its host by probing with its ovipositor into the medium in which the host lives. The host has a genotype which burrows deeply and reproduces slowly and one which burrows less deeply and lays more eggs. The parasite has a genotype which searches quickly but ineffectively for deep dwellers and one which searches carefully but more slowly. The population dynamics for such a host parasite interaction with discrete generations can be chaotic [3]. In this case, the gene frequencies may also behave in a chaotic manner. The scenario is the following: if the host population starts with few individuals, the shallow dwellers predominate originally. This allows the fast searching parasites to increase, allowing the deep dwelling hosts to become relatively more abundant. As the population of fast searching parasites crashes for lack of shallow dwelling hosts, the careful probers become rela-

tively abundant. Finally, all populations return to a low
level. The cycle may or may not repeat itself in a regular
fashion, depending upon the parameters in the model.

If one compares this scenario with the urn models of
classical population genetics, then it is evident that something quite different is taking place here. In equilibrium
urn models, the number of balls which are returned to the
urn is fixed, and different genotypes compete for increasing
their share of the balls. In the chaotic model, the number
of balls returned to the urn depends upon the total genetic
composition and sizes of the populations. The dynamics keep
the populations and their genetic structures within certain
limits so that the populations remain polymorphic. Moreover, the gene frequencies continue to fluctuate. "Natural
selection" is operating within the system. Yet, it is not
clear whether one can give a definition of fitness for which
it is reasonable to say that fitness is maximized. Nicholson'
experiments on blowflies in the laboratory showed that the
impact of evolutionary changes on the population dynamics of
the blowflies depended upon the nature of the competitive
effects causing the evolution. In some situations, the blowflies evolved so that their numbers clearly declined as a
result of the evolutionary change. How does one interpret
natural selection for populations with chaotic dynamics?

There is one final ramification of this question to
consider. The concept of r versus K selection [25] is

one that has attracted much attention in ecology. Within the
context of age structured populations models incorporating
genetic variability, it is possible to simulate the evolution of life history strategies. [6] With assumptions on
the nature and extent of the genetic variability of vital
parameters for populations, one can watch model populations
evolve. <u>Can simulations of this sort shed light on the
evolution of life history strategies?</u> In particular, can
one reproduce the qualitative results of Nicholson's evolution experiments.

We are left in a position of being cautious assessing
the future of mathematical models in ecology. There may well
be limits on the extent to which experimental data can be
readily understood in terms of detailed models. The amount
of data necessary for comparison with models is formidable.
Even then, separating that part of the data which should be
matched by a deterministic model from the part due to random
influences may be impossible in principle. The sensitivity
of models to changes in its parameters may be difficult to
assess and may depend upon the details of how each model is
constructed. This suggests that general ecological principles
of wide applicability may be rare. Extensive experience with
many complicated models may be a prerequisite for a reasonable quantitative understanding of the laws of population
dynamics. Rather than general principles of universal
validity, we may be left with techniques of simulation which

can be applied only with great effort to specific situations.

IV AFTER THOUGHT.   We have looked at two modern mathematical
theories which seem to have implications for biology in this
essay.   Both catastrophe theory and dynamical systems are
theories developed with the idea of genericity as a central
concern.   In looking at ways in which theories can be applied
to biology, a large number of interesting mathematical
questions were raised.   Very often efforts in mathematical
biology depend upon existing mathematics and try to make the
biology fit these existing theories.   But the biology is the
real world, and the models must fit it rather than the re-
verse.   This places a demand on the biomathematician for
new mathematics, appropriate to the biological problems.   It
is a challenging task, but a lot of good mathematics has
started with attempts to understand real phenomena and has
then enriched our understanding of those phenomena.

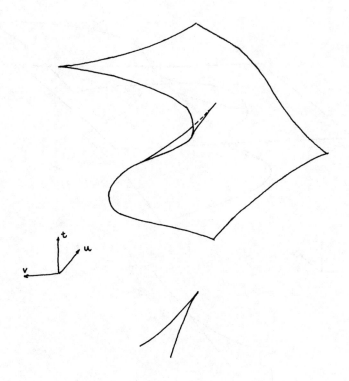

Figure 1

The cusp surface $t^3 + ut + v = 0$ and the projection of its fold into the u,v plane.

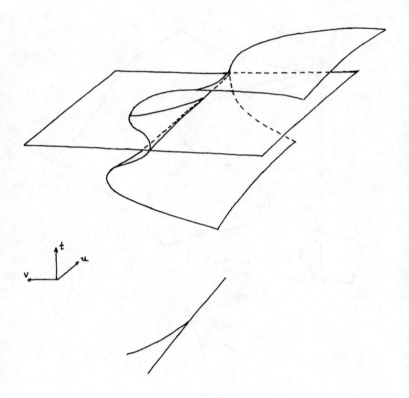

Figure 2

The surface of potential equilibria and the unfolding
of $f(x) = x^6$ in the space of even functions.

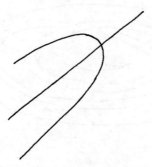

Figure 3a

Two dimensional section of cusp through the cusp point.

Figure 3b

Generic two dimensional section of cusp near the cusp
point.

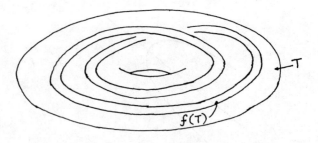

Figure 4a

A map of a solid torus into itself having a strange
attractor.

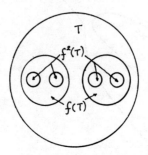

Figure 4b

A cross section of the torus.

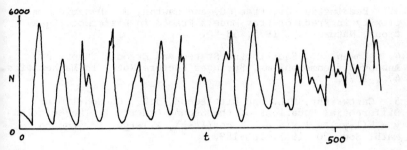

Figure 5

Data from one of Nicholson's laboratory experiments with
blowflies. Number of adults (N) vs. time in days (t).
Redrawn from Nicholson [31] .

BIBLIOGRAPHY

1.   Arnold, V.I., Critical Points of Smooth Functions, Proceedings International Congress of Mathematicians, v. 1,
Vancouver, B.C., 1974, 19-40.

2.   Baas, N., Structural Stability of Composed Mappings.

3.   Beddington, J., Free, C. and Lawton, J., Dynamic Complexity in Predator-Prey Models Framed in Difference Equations, Nature 255, 1975, 58-60.

4.   Bowen, R., Equilibrium States and Ergodic Theory of
Anosov Diffeomorphisms, Springer Lecture Notes in Mathematics,
470, 1975.

5.   Cartwright, M.L. and Littlewood, J.E., On Non-linear
Differential Equations of the Second Order: I. The Equation
$\ddot{y} + k(1-y^2) \dot{y} + y = b \lambda k \cos(\lambda t + a)$, k large. J. London
Math. Soc. 20, 1945, 180-189.

6.   Charlesworth, B., Natural Selection in Age-Structured
Populations, Some Mathematical Questions in Biology VII,
1976, 69-88.

7.   Chow, S., Hale, J. and Mallet-Paret, J., Applications
of Generic Bifurcation II Arch. Rational Mech. Anal., 62,
1976, 209-235.

8.   Croll, J., Is Catastrophe Theory Dangerous? New
Scientist, 70, 1976, 630-632.

9.   Crow, J.F. and Kimura, M., Introduction to Population
Genetics Theory, Harper and Row, New York, 1970.

10.   Gollub, J. and Swinney, H., Onset of Turbulence in a
Rotating Fluid, Physical Review Letters, 35, 1975, 927-930.

11.   Guckenheimer, J., Review of Stabilité Structurelle et
Morphogenese, Bulletin of Am. Math Soc., 79, 1973, 878-890.

12.   Guckenheimer, J., Caustics, Global Analysis and Its
Applications, v. II, International Atomic Energy Agency,
1974, 281-289.

13.   Guckenheimer, J., Catastrophes and Partial Differential
Equations, Annales de I'Institut Fourier, 23, 1973, 31-59.

14.   Guckenheimer, J., Caustics and Non-degenerate Hamilton-
ian, Topology, 13, 1974, 127-133.

15. Guckenheimer, J., Shocks and Rarefactions in Two Space
Dimensions, Arch. Rat. Mech. Anal. 59, 1975, 281-291.

16. Guckenheimer, J., Cusps of Zeeman's Catastrophe Machine,
Topology, to appear.

17. Guckenheimer, J., Bifurcations of Maps of the Unit Interval, to appear.

18. Guckenheimer, J., Oster, G., and Ipaktchi, A., Dynamics
of Density Dependent Population Models, J. of Mathematical
Biology, to appear.

19. Hassell, M., Lawton, J. and May, R., Patterns of Dynamical Behavior in Single-species Populations, J. Animal
Ecology, 45, 1976 471-86.

20. Holmes, P.J. and Rand, D.A., Bifurcations of the Forced
van der Pol Oscillator, preprint, Southamption.

21. Holmes, P.J. and Rand, D.A., The Bifurcations of Duffing's
Equation: An Application of Catastrophe Theory, Journal of
Sound and Vibration, 44, 1976, 237-253.

22. Jenny, H., Kymatik, Basilius, Basel, 1967.

23. Lewontin, R.C. and Hubby, J.L., A Molecular Approach to
the Study of Generic Heterozygosity in Natural Populations. II.
Genetics, 54, 1966, 595-609.

24. Levinson, N., A Second Order Differential Equation with
Singular Solutions, Annals of Math, 50, 1949, 127-153.

25. MacMarthur, R.H. and Wilson, E.O., The Theory of Island
Biogeography, Princeton Univ. Press, 1967.

26. Mather, J.N., Stability of C^∞ Mappings
I.   Ann. of Math, 87, 1968, 89-104
II.   Ann. of Math, 89, 1969, 254-291
III. Publ. Math. I.H.E.S., 35, 1968, 127-156
IV.   Publ. Math. I.H.E.S., 37, 1969, 223-248
V.   Advances in Math., 4, 1970, 301-335
VI.   Springer Lecture Notes in Math, 192, 1971, 207-254.

27. May, R.M., Stability and Complexity in Model Ecosystems
Princeton University Press 1974.

28. May, R.M., Simple Mathematical Models with very Complicated Dynamics, Nature 261, 1976, 459-466.

29. May, R. and Oster, G., Bifurcations and Dynamic Complexity in Simple Ecological Models, American Naturalist, 110, 1976, 573-599.

30. Murray, J., Genetic Diversity and Natural Selection Oliver and Boyd, Edinburgh, 1972.

31. Nicholson, A.J., The Self-Adjustment of Populations to Change Cold Spring Harbor Symp. Quant. Biol XXII, 1957, 153-173.

32. Newhouse, S., Diffeomorphisms with Infinitely Many Sinks, Topology, 12, 1974, 9-18.

33. Nicholson, A.J., The Role of Population Dynamics in Natural Selection, Evolution after Darwin v.I, ed. Sol Tax, U. of Chicago Press, 477-522, 1960.

34. Oster, G., Internal Variables in Population Dynamics, Some Questions in Mathematical Biology VII, 1976, 37-68.

35. Plikin, P., Sources and Sinks in Axiom A Diffeomorphisms on Surfaces, Mat. Sbornik, 94, 1974, 243-264.

36. Rhee, H., Aris, R. and Amundson, N., On the Theory of Multicomponent Chromatography, Philosophical Transactions of the Royal Society of London 267, 1970, 419-455.

37. Smale, S., Differentiable Dynamical Systems, Bull, Am. Math. Soc. 73, 1967, 747-817.

38. Sussmann, H.J., Catastrophe Theory: A Preliminary Critical Study, preprint to appear in Proceedings of 1976 Biennial Meeting of Philosophy of Science Association.

39. Takens, T., Forced Oscillations and Bifurcations, Applications of Global Analysis, Communications of Maths. Institute, Rijksuniversiteit, Utrecht, 3, 1974, 1-59.

40. Thom, R., Structural Stability and Morphogenesis, W.A. Benjamin, Inc., Reading, Mass. 1975.

41. Thom, R., A Global Dynamical Scheme for Vetebrate Embryology, Some Mathematical Questions in Biology VI, 1973, 3-45.

42. Toulouse, G. and Kléman, M., Principles of a Classification of Defects in Ordered Media, LeJournal de Physique, 37, 1976, L149-L151.

43. Turing, A.M., Chemical Basis of Morphogenesis Phil. Trans, Roy. Soc. B237, 1952, 37-72.

44. White, E.G. and Huffaker, C.B., Regulatory processes and population cyclicity in laboratory populations of Anagasta kühniella (Zeller) (Lepidoptera: Phycitidae). II. Parasitism, predation, competition and protective cover, Res. Pop. Ecol. 11, 1969, 150-185.

45. Whithman, G.B., Linear and Nonlinear Waves, Wiley, New York, 1974.

46. Whitham, G.B., A General Approach to Linear and Nonlinear Dispersive Waves Using a Lagrangian J. Fluid Mech. 22, 1965, 273-282.

47. Winfree, A.T., Spiral Waves of Chemical Activity, Science, 175, 1972, 175-177 and cover photograph.

48. Zeeman, E.C., A Catastrophe Machine, Towards a Theoretical Biology, 4, ed. C.H. Waddinton, Edinburgh University Press, 1972, 276-282.

49. Zeeman, E.C. in Structural Stability, the Theory of Catastrophes, and Applications in the Sciences. Edited by P. Hilton. Lecture Notes in Mathematics 525, Springer-Verlag, Berlin. 1976.

JOHN GUCKENHEIMER

UNIVERSITY OF CALIFORNIA,

SANTA CRUZ

Lectures on Mathematics in the Life Sciences
Volume 10, 1978

QUALITATIVE EFFECTS OF DIFFUSION IN CHEMICAL SYSTEMS

J. F. G. Auchmuty

I. Introduction

There has recently been a great revival of interest in
nonlinear diffusion equations and especially in so-called
"reaction-diffusion" or "interaction-diffusion" equations.
Such equations describe systems where the components undergo
mutual diffusion and interact according to some (usually
nonlinear) rate law. These systems have been studied as
simple models of various topics in ecology, genetics and
morphogenesis (see Glansdorff and Prigogine (1971), Levin
(1976) and Nicolis and Prigogine (1977)). Reaction diffusion
equations have also been extensively studied by chemists and
chemical engineers (see Aris (1975), Tyson (1976)).

In this paper some aspects of the mathematical analysis
of these equations will be discussed and certain qualitative
features of the solutions will be described.

As will be seen, proving the existence, uniqueness and
regularity of solutions of the equations is not the major
mathematical task in analyzing these systems. Rather the
most interesting problems are those of describing their
qualitative behaviour. The solutions of reaction-diffusion
patterns may display a variety of striking phenomena which
one would like to analyze. These have included rotating and

49

scroll-like solutions (Winfree (1974 a,b) and the attractive patterns as well as the oscillating and rotating solutions found by Herschkowitz-Kaufman (1975) and Erneux and Herschkowitz-Kaufman (1977). The solutions described in the last two papers were all obtained by solving the same equations, changing only the domains, the boundary conditions or one of the parameters in the system.

From a mathematical point of view, one would like to characterize the equations that may arise in reaction-diffusion modelling and establish a general framework within which the equations may be analyzed. One would also like to know how the solutions depend on various parameters, how sensitive they are to certain types of perturbations and what the asymptotic behaviour of the system is, either as time increases, or as various parameters tend to limiting values. In this last category one is particularly interested in comparing the behaviour of systems with and without diffusion.

The variables in these equations usually represent quantities, such as temperatures and concentrations, which should be non-negative. Consequently special attention will be paid to the nonnegativity of solutions. The systems will always be assumed to occupy a bounded region in space and be subject to boundary conditions. There is a considerable difference between the theories for nonlinear diffusion equations defined on bounded, as against unbounded, domains

in space. Thus the methods used and the results obtained
will be quite different to those described recently by
Aronson and Weinberger (1975) or by Howard and Kopell (1973)
who treated problems on unbounded domains.

The plan of the paper is as follows. In section 2 we
treat some aspects of chemical reactions without diffusion.
In section 3 the theory of reaction-diffusion equations is
introduced and some effects of diffusion are described. A
number of examples of these systems are mentioned in Section
4. Sections 5 and 6 treat steady-state and time-periodic
solutions respectively and finally we conclude with some
comments on reaction-diffusion modelling and catastrophe
theory.

II. Chemical Kinetics

The evolution of a chemical system is determined by the
rate functions for the system so, in this section, we shall
investigate various assumptions about the rate functions and
their consequences. Throughout this paper these functions
will be assumed known. For an introduction to the
foundations of chemical kinetics see Aris (1965), (1968),
Gavalas (1968), Horn and Jackson (1972), Krambeck (1970)
and Oster and Perelson (1974) amongst others.

Consider a system of r chemical reactions involving
K chemical species $A_1, A_2, \ldots, A_K$. This may be written

$$\sum_{j=1}^{K} \nu_{ij} A_j = 0 \qquad 1 \leq i \leq r \qquad (2.1)$$

where $\nu = (\nu_{ij})$ is the stoichiometric matrix of the system.
If the system is "well-stirred" the state of the chemical
system at any time t is usually described by a vector
$u(t)$ whose components are the concentrations of the
chemical species and, perhaps, certain thermodynamic
variables such as temperature, pressure or volume. Assume
that one can choose units so that this vector only has
non-negative components.

Here we shall consider chemical systems whose evolution
is given by the solutions of the ordinary differential
equation

$$\frac{du_i}{dt} = f_i (u_1, u_2, \ldots, u_m) \quad 1 \leq i \leq m \qquad (2.2)$$

subject to given initial conditions

$$u_i(0) = u_{0i} \qquad\qquad 1 \leq i \leq m \qquad (2.3) \ .$$

The functions f_i are called the rate functions and need
only be defined for non-negative values of the u_i . One
has $K \leq m$. Let $\mathbb{R}^m$ denote the usual m-dimensional real
vector space and $\mathbb{P}^m$ be the nonnegative cone in $\mathbb{R}^m$,

$$\mathbb{P}^m = \{v \in \mathbb{R}^m : v_i \geq 0 \ \text{ for } \ 1 \leq i \leq m\} \ .$$

Let $f : \mathbb{P}^m \to \mathbb{R}^m$ be the function whose components are

f_i . It will be assumed to satisfy

    (F1) :   f is Lipschitz continuous on each bounded

           subset of $\mathbb{P}^m$.

    (F2) :   if v lies in the boundary $\partial\mathbb{P}^m$ of $\mathbb{P}^m$,

           then $f_i(v) \geq 0$ for those i such that $v_i = 0$.

When f obeys these two conditions and $u_0 \in \mathbb{P}^m$, then

a unique solution of (2.2)-(2.3) exists, at least locally

in time. Moreover Nagumo (1942) has shown that (F2) is

necessary and sufficient for such a solution to remain in

$\mathbb{P}^m$ during its interval of existence. Condition (F2) has

the geometrical interpretation that, on the boundary of the

positive cone, the vector field defined by f should not

point outwards.

If the maximal time of existence of the solution of

(2.2)-(2.3) is $T(u_0)$, then the trajectory of the solution

is the set $\{u(t) : 0 \leq t < T(u_0)\} \subset \mathbb{R}^m$.

A vector v in $\mathbb{P}^m$ is called a steady-state (or

equilibrium or time-independent) solution of (2.2) if the

solution of (2.2) subject to $u_i(0) = v_i$ for $1 \leq i \leq m$,

is $u(t) \equiv v$, $0 \leq t < \infty$.

Obviously if v is a steady-state solution then

$$f_i(v) = 0 \qquad\qquad 1 \leq i \leq m. \qquad\qquad (2.4)$$

Very often there is further structure on these systems.

If the system is closed, the number of atoms of each element is conserved and one obtains a number of linear conservation laws. In such a case there exists an $\ell \times m$ matrix $\varepsilon = (\varepsilon_{ij})$ such that

$$(F3): \sum_{k=1}^{m} \varepsilon_{ik} f_k(u) = 0 \quad \text{for} \quad 1 \le i \le \ell, \quad \text{all} \quad u \in \mathbb{P}^m.$$

Equation (2.2) yields

$$\frac{d}{dt} \left(\sum_{k=1}^{m} \varepsilon_{ik} u_k(t) \right) = 0 \quad 1 \le i \le \ell. \tag{2.5}$$

Thus

$$\sum_{k=1}^{m} \varepsilon_{ik} u_k(t) = \sum_{k=1}^{m} \varepsilon_{ik} u_{k0} \quad 1 \le i \le \ell, \tag{2.6}$$
$$0 \le t < T(u_0)$$

When the system defined by $(2.2)-(2.3)$ obeys $(F1)-(F3)$ and $u_0 \in \mathbb{P}^m$, define

$$C(u_0) = \{v \in \mathbb{P}^m : \sum_{k=1}^{m} \varepsilon_{ik}(v_k - u_{k0}) = 0 \quad 1 \le i \le \ell\}.$$

Then the trajectory of any solution is a subset of $C(u_0)$ and $C(u_0)$ is a closed convex subset of $\mathbb{P}^m$ called the reaction simplex.

Another important feature of many chemical systems is that they have natural Lyapunov functions. These natural Lyapunov functions arise from thermodynamic considerations. A function $V : \mathbb{P}^m \to \mathbb{R}$ is said to be a Lyapunov function for the equation (2.2) provided

(V_1): V is continuous on $\mathbb{P}^m$ and continuously differentiable on the interior $\overset{\circ}{\mathbb{P}}{}^m$ of $\mathbb{P}^m$, and (V_2): $\frac{d}{dt} (V(u(t))) \leq 0$ along any trajectory of (2.2)

with strict inequality holding whenever $u(t)$

is not a steady-state solution of (2.2).

A point w in $\overset{\circ}{\mathbb{P}}^m$ is a critical point of V if

grad $V(w) = 0$.

The natural Lyapunov functions arising in chemical

systems often obey a further condition, namely

(V_3): A point w is a critical point of V if and

only if w is a steady state solution of (2.2).

A function which obeys $(V_1)-(V_3)$ will be called a

global Lyapunov function. Condition (V_3) is the state-

ment that the thermodynamic equilibria of a system are the

same as the kinetic equilibria (see Glansdorf and Prigogine

(1971), Othmer (1976a)). Examples of such natural Lyapunov

functions in chemical systems include the Gibbs' free

energy in a closed, isothermal, isobaric system or the

negative entropy in a closed, adiabatic, isochoric system.

The role of Lyapunov functions in chemical systems has been

discussed by Wei (1962), Gavalas (1968) and Krambeck (1970).

To illustrate the use of these concepts in obtaining

results on the qualitative behaviour of chemical systems we

shall prove the following extension of a result in Wallwork

and Perelson (1976).

A function $V : \mathbb{P}^m \to \mathbb{R}$ is said to be strictly convex

on a convex subset K of $\mathbb{P}^m$ if

$$V(\alpha u + (1-\alpha)v) < \alpha V(u) + (1-\alpha)V(v) \quad \text{for } 0<\alpha<1 \ ; \ u,v \in K \ .$$

<u>Theorem</u> 1. Suppose (2.2)-(2.3) describes the evolution of a
chemical system which obeys (F1)-(F3) and where

    (i)   each reaction simplex $C(u_0)$ is bounded, and

    (ii) there is a global Lyapunov function which is
        strictly convex on $C(u_0)$.

Then the solution of (2.2)-(2.3) is defined for $0 \le t < \infty$,
there is a unique steady-state solution w of (2.2) in
$C(u_0)$ and $\lim_{t \to \infty} u(t) = w$. w is the unique minimum of V
on $C(u_0)$.

<u>Proof.</u> V is a strictly convex function on the bounded
closed set $C(u_0)$ so it attains a unique minimum w there.
Let $v(t)$ be the solution of (2.2) subject to $v(0) = w$.
Then, from V_2 , $V(v(t))$ cannot increase with time so
$v(t) \equiv w$ for $t > 0$, so w is a steady-state solution lying
in $C(u_0)$.

    Let w_1 be another steady-state solution in $C(u_0)$.
Then, from V_2 and V_3 , w_1 must also be a critical point
of V . But a strictly convex function has a unique
critical point so $w = w_1$.

    Let $u(t)$ be the solution of (2.2)-(2.3). Since $u(t)$
remains in the bounded set $C(u_0)$ it is defined for
$0 \le t < \infty$. Let $v_0 (\ne w)$ be a limit point of the trajectory
of $u(t)$ as $t \to \infty$. Then there exists a sequence $\{t_n\}$

with $t_n \to \infty$ such that $u(t_n) \to v_0$. Also $V(u(t)) \geq V(v_0)$
for all $t > 0$ as V is continuous and non-increasing
along trajectories. Let $v(t)$ be the solution of (2.2)
subject to $v(0) = v_0$. Then, for any $\tau > 0$, $V(u(t_n + \tau))$
converges to $V(v(\tau))$ which is less than $V(v_0)$ from
V_2 and V_3 . This contradicts the fact that $V(u(t)) \geq V(v_0)$
so v_0 must equal w and the theorem holds.

To apply this to the situation studied by Wallwork and
Perelson (1976), one notes that a closed, isobaric,
isothermal, single-phase system obeys the conditions of this
theorem with the Gibbs' free energy being the Lyapunov
function. Henry's law then implies that the minimum of the
Gibbs' free energy is attained at an interior point of
$C(u_0)$. Thus their Theorem 2 holds without the hypothesis
of "ω-stability".

Similar results have been described in Krambeck (1970)
and Horn & Jackson (1972).

Note that when f is not required to obey (F3), but
there is a strictly convex Lyapunov function for the system,
one has similar results.

Theorem 2. Suppose $(2.2)-(2.3)$ describes the evolution of a
chemical system whose rate functions obey $(F1)-(F2)$.
Suppose the system has a global Lyapunov function V which
is strictly convex on $\mathbb{P}^m$ and there exists a constant M

58   J. F. G. AUCHMUTY

such that $V_M = \{u \in \mathbb{P}^m : V(u) \le M\}$ is non-empty and bounded.
Then there is a unique steady-state solution w of (2.2)-
(2.3) in $\mathbb{P}^m$ and if $u_0 \in V_M$, then the solution of (2.2)-
(2.3) is defined for $0 \le t < \infty$ and obeys $\lim_{t \to \infty} u(t) = w$. w
is the unique minimum of V in $\mathbb{P}^m$.

<u>Proof</u>. Since V decreases along trajectories, any
trajectory that starts in V_M must remain there. So such
a solution of (2.2)-(2.3) exists for all time. Also V
attains its minimum on V_M and the minimum is unique since
V is strictly convex. The remaining parts of the theorem
are proven just as in Theorem 1.

The condition on V_M is satisfied, for example, by
the Gibbs' free energy function for an isothermal, isobaric
system, as it obeys $\lim_{\|u\| \to \infty} V(u) = +\infty$. Later we shall use
the following condition:

(V_4): For all real M, $V_M = \{u \in \mathbb{P}^m : V(u) \le M\}$ is
bounded.

Sometimes one might try to analyze chemical systems for
which (F2) doesn't hold but where one is still only
interested in non-negative solutions. In such a case one
might have exhaustion of some of the chemicals. Then one
usually redefines the system of equations so that

$$\frac{du_i}{dt}(t) = \begin{cases} f_i(u(t)) & \text{if } u_i > 0 \\ \max(0, f_i(u(t))) & \text{if } u_i = 0 \end{cases} \quad 1 \le i \le m.$$

This is called an evolution inequality and one can show that the solution of this lies in $\mathbb{P}^m$ during its interval of existence. For an introduction to the theory of such problems see Duvaut and Lions (1972) or Lions (1969).

III. Diffusion in Chemical Systems

Quite often, and especially in biological systems, one does not have "well-stirred" chemical reactions and one is interested in describing and analyzing the spatial behaviour of the system. There are a number of transport mechanisms which may induce spatial inhomogenieties in chemical systems. They include convection and fluid motions as well as diffusion. Here, however, attention will be confined to diffusive effects because they appear to be the simplest to analyze as well as being quite common.

A very good introduction to reaction-diffusion systems is given in Chapter 1 of Volume 1 of Aris (1975), where the physical and chemical bases are thoroughly treated. Much of the following analysis carries over to other interaction-diffusion systems arising in ecological or genetic problems where one has similar equations with the interactions also obeying conditions similar to (F1)-(F2).

Consider a chemical system in a bounded open set Ω in $\mathbb{R}^n$, $1 \le n \le 3$. When $n > 1$, assume the boundary $\partial\Omega$ of Ω is smooth and write $\bar{\Omega} = \Omega \cup \partial\Omega$. Let the state of the system be described by a vector valued function $u(x,t)$ taking values in $\mathbb{P}^m$, where the components of $u(x,t)$ represent the concentrations, temperature etc. at the point (x,t).

Assume that the diffusion is homogeneous and is governed by Fick's law, so the diffusive flux j_i of u_i is given by

$$j_i = -D_i(x,u)\,\text{grad}\,u_i \qquad 1 \le i \le m \ .$$

Henceforth we shall assume that for each $1 \le i \le m$, there exists a $\delta_i > 0$ such that $D_i(x,u) \ge \delta_i > 0$ for all $x \in \bar{\Omega}$, $u \in \mathbb{P}^m$. It will also be assumed that each $D_i(x,u)$ is continuously differentiable on $\Omega \times \mathbb{P}^m$, and for each u in $\mathbb{P}^m$, $D_i(x,u)$ is bounded above on $\bar{\Omega}$.

Let the reaction rates be described by rate functions f_i as in the last section. Then the equations for coupled reaction and diffusion are

$$\frac{\partial u_i}{\partial t} = \sum_{j=1}^{n} \frac{\partial}{\partial x_j}\left(D_i(x,u)\,\frac{\partial u_i}{\partial x_j}\right) + f_i(x,u) \quad \text{on } \Omega \times (0,T), \quad (3.1)$$
$$1 \le i \le m.$$

This is to be solved subject to given initial and boundary conditions:

$$u_i(x,0) = u_{0i}(x) \ge 0 \qquad x \text{ in } \bar{\Omega}, \quad 1 \le i \le m \qquad (3.2)$$

$$\alpha_i(x)\,\frac{\partial u_i}{\partial \nu}(x,t) + \beta_i(x)u_i(x,t) = \gamma_i(x), (x,t) \in \partial\Omega \times (0,T)$$
$$1 \le i \le m. \qquad (3.3)$$

Here α_i, β_i and γ_i are non-negative, continuous functions on $\partial\Omega$ with $\alpha_i^2(x) + \beta_i^2(x) \neq 0$ for all x in $\partial\Omega$. $\frac{\partial}{\partial\nu}$ is the directional derivative in the direction of the outward normal at a point x on the boundary.

The most common boundary conditions are either "no flux" or Neumann boundary conditions

$$\frac{\partial u_i}{\partial\nu}(x,t) = 0 \quad \text{for} \quad (x,t) \in \partial\Omega \times (0,T), 1\leq i\leq m, \quad (3.4)$$

or "bath" or Dirichlet conditions

$$u_i(x,t) = \gamma_i \geq 0 \text{ for } (x,t) \text{ in } \partial\Omega \times (0,T), 1 \leq i \leq m. \quad (3.5)$$

The following conditions on the functions $f : \bar{\Omega} \times \mathbb{P}^m \to \mathbb{R}^m$ will be used;

($\mathcal{F}1$): f is continuous and $f(x,u)$ is Lipschitz continuous in u on each bounded subset of $\mathbb{P}^m$, uniformly for x in $\bar{\Omega}$.

($\mathcal{F}2$): when v lies in the boundary $\partial\mathbb{P}^m$ of $\mathbb{P}^m$, then $f_i(x,v) \geq 0$ for all x in $\bar{\Omega}$ and for those i for which $v_i = 0$.

When $f(x,u) = g(u)$ is independent of X, conditions ($\mathcal{F}1$) and ($\mathcal{F}2$) reduce to conditions (F1) and (F2) of the last section. In this case, if the initial conditions are also independent of x, so that $u_i(x,0) \equiv u_{i0}$, then the solutions of (2.1)-(2.2) are spatially uniform solutions of (3.1)-(3.2) and (3.4). Hence any solution of the ordinary differential equations is also a solution of the reaction-

diffusion system with uniform initial distribution and no
flux boundary conditions.

The following condition which is a slight strengthen-
ing of ($2) will be used

($2'): when $(y,v) \in \Omega \times \partial \mathbb{P}^m$, there exists neighborhoods

V of v and W of y such that (x,u) in

$W \times (V \cap \mathbb{P}^m)$ implies that $f_i(x,u) \geq 0$ for those

i for which $v_i = 0$.

Let $u : \bar{\Omega} \times [0,T) \to \mathbb{P}^m$ be a continuous function whose
components u_i, $1 \leq i \leq m$ are twice continuous differentiable in x_j, $1 \leq j \leq n$ and continuously differentiable in t at
each point (x,t) in $\Omega \times (0,T)$. Such a function is a
classical solution of (3.1)-(3.3) on $\Omega \times (0,T)$ provided
the components $u_i(x,t)$, $1 \leq i \leq m$ have a normal derivative
at each point (x,t) on $\partial \Omega \times (0,T)$ and equations (3.1)-(3.3)
are obeyed.

When f obeys ($1), and T is small enough, there
is a unique classical solution of (3.1)-(3.3). To prove
that the solutions exist for all time one must obtain
various a priori bounds on the solutions. This has been
done if f obeys certain extra conditions or for
particular model systems (see Henry (1976), Friedman (1965),
Diekmann and Temme (1976)). Henceforth we shall always
assume our systems are such that classical solutions exist
and are unique.

Various authors have proven results on the non-negativity, or on the existence of invariant sets for equations (3.1)-(3.3) under various extra hypotheses. A subset S of $\mathbb{P}^m$ is said to be an invariant set for the solutions of (3.1)-(3.3) if every classical solution of (3.1)-(3.3) subject to $u_0(x) \in S$ for all x in $\bar{\Omega}$ obeys $u(x,t) \in S$ for all (x,t) in $\bar{\Omega} \times [0,T)$.

Martin (1973) and Redheffer and Walter (1975) have studied abstract problems of this form. Weinberger (1975) has obtained results when the diffusion terms are the same for each component. Chueh, Conley and Smoller (1977) have obtained some very general results on invariant sets. Many other authors have proved such results for particular systems. For the problem (3.1)-(3.3), one can use results in Auchmuty (1977) to prove the following.

<u>Theorem 3</u>. Let u be a classical solution on $\Omega \times [0,T)$ of the system (3.1)-(3.3) as above with $\gamma_i(x) \geq \delta > 0$ for all x in $\partial\Omega$, $1 \leq i \leq m$. Assume f obeys (\mathcal{F}1) and (\mathcal{F}2') and $u_{0i}(x) > 0$ for all $1 \leq i \leq m$, x in $\bar{\Omega}$, then $u_i(x,t) > 0$ for all (x,t) in $\bar{\Omega} \times (0,T)$ and $1 \leq i \leq m$.

To obtain this result from the proof of Theorem 1 in Auchmuty (1976) one need only note that we have made sufficiently many assumptions on the diffusion coefficients $D_i(x,u)$ that lemma 1 in that paper may still be applied, and otherwise the argument is the same.

Essentially this theorem says that if the initial conditions are positive and the other conditions of the theorem hold then the solution remains positive during its whole interval of existence.

Again, just as for ordinary chemical kinetics, if f does not obey a condition such as (32) or (32') one sometimes has to replace the equations (3.1)-(3.3) by a variational inequality to ensure the non-negativity of the solutions. This was, in effect, done by Turing (1952) in his famous paper on morphogenesis.

A function $\mathbf{v} : \bar{\Omega} \to \mathbb{P}^m$ will be called a steady state solution of (3.1), (3.3) provided that $u_i(x,t) \equiv v_i(x)$ is a solution of (3.1), (3.3). Thus v obeys (3.3) on $\partial\Omega$ and

$$\sum_{j=1}^{n} \frac{\partial}{\partial x_j} \left(D_i(x,v) \frac{\partial v_i}{\partial x_j} \right) + f_i(x,\mathbf{v}) = 0 \quad \text{on } \Omega , \quad (3.6)$$

$$1 \le i \le m.$$

Certain properties of the steady-state solutions will be described in section 6.

Next, consider the effect of introducing diffusion to systems whose kinetics yielded a certain number of linear conservation laws or had a Lyapunov function.

When u is a classical solution of (3.1)-(3.3) on $\bar{\Omega} \times [0,T)$, define

$$\bar{u}_i(t) = \frac{1}{|\Omega|} \int_{\Omega} u_i(x,t) \, dx \quad (3.7)$$

where $|\Omega| = \int_{\Omega} dx$ is the volume of Ω .

Suppose $f_i(x,u) = f_i(u)$, $1 \le i \le m$, are independent of x and obey (F3). Then from (3.1), upon integrating by parts once,

$$\frac{d}{dt}\left(\sum_{k=1}^{m} \epsilon_{ik}\bar{u}_k(t)\right) = \sum_{k=1}^{m} \frac{\epsilon_{ik}}{|\Omega|} \int_{\partial\Omega} D_k(x,u)\frac{\partial u_k}{\partial\nu}\, d\sigma \ , \quad 1 \le i \le \ell \ .$$

Here $d\sigma$ is an element of surface area on $\partial\Omega$.

When no flux boundary conditions (3.4) are imposed, this yields

$$\sum_{k=1}^{m} \epsilon_{ik}\bar{u}_k(t) = \sum_{k=1}^{m} \epsilon_{ik}\bar{u}_k(0) \ , \qquad \begin{matrix} 0 \le t < T \\ 1 \le i \le \ell. \end{matrix} \qquad (3.8)$$

In this case one again has ℓ linear conservation laws, and there is an analogue of the reaction simplex in function space. In general one sees that if other boundary conditions hold, then the conservation laws are broken.

Now consider a chemical system whose kinetics admit the existence of a global Lyapunov function V obeying V_1, V_2 and V_3 . Suppose the diffusion coefficients D_i are positive and independent of u . Define

$$\mathcal{V}u(t) = \frac{1}{2}\sum_{i=1}^{m}\sum_{j=1}^{n}\int_{\Omega} D_i(x)\left(\frac{\partial u_i}{\partial x_j}\right)^2 dx + \int_{\Omega} V(u(x,t))\,dx. \quad (3.9)$$

When $u(x,t)$ is a classical solution of (3.1)-(3.2) subject to (3.4),

$$\frac{d}{dt}\,\mathcal{V}u(t) = -\sum_{i=1}^{m}\int_{\Omega}\left[\sum_{j=1}^{n}\frac{\partial}{\partial x_j}\left(D_i(x)\frac{\partial u_i}{\partial x_j}\right) - \frac{\partial V}{\partial u_i}\right]\frac{\partial u_i}{\partial t}\, dx$$

$$= \sum_{i=1}^{m}\int_{\Omega}\left[f_i(u) - \frac{\partial u_i}{\partial t} + \frac{\partial V}{\partial u_i}\right]\frac{\partial u_i}{\partial t}\, dx \ .$$

If the kinetics are obtained from a gradient system, namely

$$(V5): \quad f_i(u) = -\frac{\partial V}{\partial u_i}(u) \qquad 1 \le i \le m, \quad u \text{ in } \mathbb{P}^m$$

one sees that

$$\frac{d}{dt}\mathbf{V}u(t) = -\int_\Omega \sum_{i=1}^m \left(\frac{\partial u_i}{\partial t}\right)^2 dx \le 0, \tag{3.10}$$

with equality holding only when

$$\frac{\partial u_i}{\partial t}(x,t) \equiv 0 \quad \text{for all } x \text{ in } \Omega, \quad 1 \le i \le m.$$

For these systems, one may use Lyapunov methods for parabolic equations, (c.f. Auchmuty (1973), Henry (1976), Dikkmann and Temme (1976)) to obtain results on the stability and the asymptotic behaviour of solutions. To do this we shall also assume that the rate functions only grow polynomially in u, and use the terminology of Auchmuty (1973).

Given u in $\mathbb{P}^m$, define $|u| = \sum_{i=1}^m |u_i|$ and

(F4): there exists q, $1 < q < \infty$ such that

$$\limsup_{|u| \to \infty} \frac{|f(u)|}{|u|^{q-1}} = M < \infty.$$

Theorem 4. Suppose $u: \bar{\Omega} \times [0,\infty) \to \mathbb{P}^m$ is a classical solution of (3.1)-(3.2) and (3.4) with f being independent of x and $D_i(x,u) \equiv D_i(x)$ being independent of u.

Let   V   be a global Lyapunov function for (2.2)-(2.3) which
obeys (V4), (V5)   and is strictly convex on $\mathbb{P}^m$, with
f   obeying (F1)-(F2) and (F4). Then there is a unique,
spatially homogeneous, steady state solution   w   with $w(x) \equiv$
$\tilde{w}$ where $\tilde{w}$ is the unique minimum on   V   on $\mathbb{P}^m$ and
$\lim\limits_{t \to \infty} u(x,t) \equiv w$ for all   x   in $\bar{\Omega}$.

<u>Proof</u>. When   f   obeys (F4), let $p = \min(q,2)$, and take
$X = W^{1,p}(\Omega) \times \ldots \times W^{1,p}(\Omega)$ where there are   m   copies of the
usual Sobolev space $W^{1,p}(\Omega)$ in   X . Let $\mathcal{P} =$
$\{u \in X : u_i(x) \geq 0$, for all   x   in $\Omega\}$. $\mathcal{P}$ is a closed,
convex set in   X . Extend $\mathcal{V}$ to   X   by defining $\mathcal{V}u = +\infty$
if $u \notin \mathcal{P}$.

   $\mathcal{V}$ is a continuous and strictly convex functional on
$\mathcal{P}$, being the sum of a convex functional and a strictly
convex functional and $w(x) \equiv \tilde{w}$ is the unique minimum and
the unique critical point of $\mathcal{V}$ on   X .

   From (3.10), one sees that $\mathcal{V}u(t) < \mathcal{V}u(0)$ if $u(x,t) \neq$
$w(x)$, so $\mathcal{V}u(t)$ is bounded along a trajectory. This
implies $u(x,t)$ is uniformly bounded in   X   as   t   varies.
Using the theory of parabolic equations in Sobolev spaces,
one finds that $\Sigma = \{u(\cdot,t) : 0 \leq t < \infty\}$ defines a pre-compact
set in   X . A function   v   in   X   is   a ω-limit point
of a solution $u(\cdot,t)$ provided there exists a sequence
$\{t_n\}$ such that $t_n \to \infty$ and $\lim\limits_{t_n \to \infty} u(\cdot,t_n) = v$ strongly in
X . Let   W   be the set of all ω-limit points of Σ . Then

W is non-empty, closed and connected in X . Now arguing
just as in Theorem 1, one can show that $W = \{w\}$ for
otherwise, if there exists $w_1 (\neq w)$ in W one gets a
contradiction.

This theorem is mostly of interest in a negative way.
It shows that there are systems whose asymptotic behaviour
is not modified by introducing diffusion, no matter what
the size of the diffusion coefficients. In the next few
sections, we shall study systems which are not gradient
systems and where the introduction of diffusion greatly
affects the behaviour of solutions.

IV.   <u>Model Systems</u>

So far most analyses of reaction-diffusion equations
have concentrated on a few specific model systems. Much of
the work has been numerical. The best known of these is
contained in Turing's (1952) famous paper in which he
proposed that combined reaction and diffusion could explain
certain aspects of morphogenesis. For recent analyses of
Turing's scheme see Smale (1974) and Erneux, Hiernaux and
Nicolis (1977).

Other models which have been extensively studied in
connection with pattern formation and morphogenesis include
those of Balslev and Degn (1975), Gierer and Meinhardt
(1972) and (1974), Othmer (1976b) and Winfree (1974).

Various other biological phenomena that have been modelled by equations of similar type include chemotaxis by Keller and Segel (1970) and Scribner, Segel and Rogers (1974), glycolytic oscillations by Goldbeter (1973) and chemical oscillations in a membrane by Caplan, Naparstek and Zabusky (1973). The books of Rashevsky (1960) and Rubinow (1975) also contains a number of nonlinear diffusion problems.

There also has been considerable study of model chemical reactors where one has flow as well as reaction and diffusion. Such work includes Amundson (1974), Aris (1975), Cohen (1972), Gavalas (1968) and Poore (1973). A number of authors have studied the effects of adding diffusion to ecological interactions including Levin (1976), Williams and Chow (1977), Segel and Jackson (1972) and Conway and Smoller (1977).

The above list is by no means complete and recently there has been an avalanche of new models. The number and variety of problems which are being modelled by reaction-diffusion models is astonishing, and many of these models have very intriguing solutions if the numerical simulations are to be believed.

For the rest of this paper, a particular model system will be used to exemplify much of the analysis. The model is the trimolecular scheme of Lefever and Prigogine (1968). It is a system which has been extensively studied both

numerically and mathematically and which can be shown to be
the simplest system having certain types of behaviour. The
scheme is

$$A \to X$$
$$2X + Y \to 3X$$
$$B + X \to Y + D$$
$$X \to E .$$

$$(4.1)$$

Here A, B, D and E are the initial and final
products whose concentrations are fixed throughout the
system. The reaction steps are all taken to be irreversible
and have rate constants equal to unity. Assuming the law
of mass-action the reaction-diffusion equations describing
this system are

$$\frac{\partial X}{\partial t} = D_1 \Delta X - (B+1)X + X^2 Y + A$$
$$\frac{\partial Y}{\partial t} = D_2 \Delta Y + BX - X^2 Y$$

$$(4.2)$$

in a domain $\Omega \times (0,T)$ subject to initial conditions

$$X(x,0) = X_0(x) \qquad Y(x,0) = Y_0(x) \quad x \in \Omega \qquad (4.3)$$

and to either $X(x,t) = A_0, \; Y(x,t) = BA_0^{-1}$ (4.4)

on $\partial\Omega \times (0,T)$.

or to $\frac{\partial X}{\partial \nu}(x,t) = \frac{\partial Y}{\partial \nu}(x,t) = 0$ (4.5)

Now we are assuming homogeneous diffusion with the
diffusion coefficients D_1, D_2 of X and Y respectively
being independent of X and Y . Δ is the Laplacian, B

is a nonnegative constant and $A = A(x)$ is a given, smooth positive function which is constant on $\partial\Omega$; $A(x) = A_0$ for x in $\partial\Omega$. Often $A(x)$ will be assumed constant throughout Ω .

This system has all the properties that one would like. The existence of nice solutions for all time and the fact that these solutions are nonnegative provided the initial conditions are nonnegative has been proved. For analyses of this system see Auchmuty and Nicolis (1975) and (1976), Diekmann and Temme (1976) and Meurant and Saut (1977). Numerical results for this system are described in Herschkowitz-Kaufman and Nicolis (1972), Herschkowitz-Kaufman (1975) and Erneux and Herschkowitz-Kaufman (1977).

Upon neglecting the diffusion terms, the system becomes

$$\frac{dX}{dt} = A - (B+1)X + X^2 Y \qquad (4.6)$$
$$\frac{dY}{dt} = BX - X^2 Y \ .$$

For any nonnegative initial conditions $(X(0),Y(0))$, the solution of this equation is defined for all time and remains nonnegative. When $B \leq 1 + A^2$, the solution tends asymptotically in time to the unique steady-state solution $X \equiv A$, $Y \equiv B/A$. When $B > 1 + A^2$ the solution tends asymptotically to a time-periodic solution or limit cycle which encloses the steady-state solution.

V.   Steady-State Solutions

The first step in analyzing the qualitative behaviour of reaction-diffusion systems is usually to try to find the steady-state solutions, particularly, the stable steady-state solutions, and to determine how they depend on the various parameters in the system. These parameters might include the diffusion coefficients, the size of the system or the boundary conditions and the rate constants or the concentrations of certain chemical species.

Let these parameters be represented by a vector μ taking values in a set $\Lambda \subseteq \mathbb{R}^k$, and suppose the equations (3.6) may be written

$$\sum_{j=1}^{n} \frac{\partial}{\partial x_j} \left(D_i(x) \frac{\partial v_i}{\partial x_j} \right) + f_i(\mu, x, \mathbf{v}) = 0 \quad \text{in } \Omega , \quad 1 \le i \le m \quad (5.1)$$

subject to

$$\alpha_i(x) \frac{\partial v_i}{\partial \nu} + \beta_i(x) v_i = \gamma_i(\mu, x) \quad \text{on } \partial\Omega , \quad 1 \le i \le m. \quad (5.2)$$

Here we are assuming that the diffusion coefficients do not depend on v and that α_i, β_i and D_i obey the same conditions as in Section 3. Also assume $f : \Lambda \times \bar{\Omega} \times \mathbb{P}^m \to \mathbb{R}^m$ and $\gamma_i : \Lambda \times \partial\Omega \to \mathbb{R}$ $1 \le i \le m$ are continuous functions and for each $\mu \in \Lambda$, $f(\mu, ., .)$ and $\gamma_i(\mu, ., .)$ obey the conditions (𝔉1) and (𝔉2) on f and the conditions on γ_i given in Section 3.

When $f_i(\mu, x, v) = g_i(\mu, v)$ is independent of x for all i then any solution w_μ of the equation

$$g(\mu, w) = 0 \qquad (5.3)$$

is a solution of (5.1). Such a solution is called a uniform
or homogeneous steady-state solution. It is a solution of
the boundary value problem when the boundary conditions are

$$\frac{\partial v_i}{\partial \nu}(x) = 0 , \quad x \text{ in } \partial\Omega , \ 1 \le i \le m , \qquad (5.4)$$

or $\qquad\qquad v_i(x) = w_{\mu_i} \quad x \text{ in } \partial\Omega . \ 1 \le i \le m , \qquad (5.5)$

or more generally if $\gamma_i(\mu, x) = \beta_i(x) w_{\mu_i}$ in (5.2).

If the boundary conditions are different to those
given in the last paragraph one can often find solutions
which are close to w_μ throughout a compact subset of Ω
but have boundary layers near the boundary of the domain.

Theorem 4 gives an example where the only steady-state
solutions of a system of the form (5.1)-(5.4) is a uniform
steady-state solution. Othmer (1976b), Conway and Smoller
(1977) and Conway, Hopf and Smoller (1977) have shown that
if the functions D_i are sufficiently large then the only
steady-states of (5.1)-(5.2) (with $\gamma_i(\mu, x) = 0$ and
$f_i(\mu, x, v) = g_i(\mu, v)$) are spatially uniform solutions.

Qualitatively this is the behaviour many people
expected of diffusion processes - namely that diffusion
would smooth out all spatial variations and one would
obtain uniform solutions. It is probably Turing's major
contribution to recognize that this need not be the case and

that diffusion can induce spatial patterning.

For given μ in Λ , a classical solution of (5.1)-(5.2) is a continuous function $u_\mu : \bar{\Omega} \to \mathbb{P}^m$ which is twice continuously differentiable on Ω and which satisfies (5.1)-(5.2).

Let $C(\bar{\Omega}; \mathbb{R}^m)$ be the Banach space of all continuous functions mapping $\bar{\Omega}$ into $\mathbb{R}^m$ with the norm

$$\|u\| = \sum_{\ell = 1}^{m} \sup_{x \in \bar{\Omega}} |u_\ell(x)| \ .$$

The set of all solutions of (5.1)-(5.2) will be denoted

$$\mathfrak{s} = \{(\mu, u_\mu) : u_\mu \text{ is a classical solution of (5.1)-(5.2)}\}.$$

Similarly let $\mathfrak{s}_0 = \{(\mu, w_\mu) : w_\mu \text{ is a solution of (5.3)}\}$.

When (5.4) or (5.5) hold, $\mathfrak{s}_0$ may be regarded as a subset of $\mathfrak{s}$. The solutions in $\mathfrak{s}$ which are not in $\mathfrak{s}_0$ are examples of "dissipative structures" (see Glansdorff and Prigogine (1971)).

To find how the solutions of (5.1)-(5.2) depend on μ one would like to find $\mathfrak{s}$ and, in particular, to describe the topological properties of $\mathfrak{s}$. A maximal, closed connected subset of $\mathfrak{s}$ (maximal with respect to set theoretic inclusion) is called a component of $\mathfrak{s}$. Let the components of $\mathfrak{s}$ be $\{\mathfrak{s}_k : k \in \mathcal{K}\}$ so that

$$\mathfrak{s} = \bigcup_{k \in \mathcal{K}} \mathfrak{s}_k \ .$$

Recently a number of authors have obtained results describing the components of $\mathcal{S}$ for various nonlinear elliptic equations. See Dancer (1973), Crandall and Rabinowitz (1970), Rabinowitz (1971) and Turner (1975). Both Dancer and Turner proved results on the existence of components of positive solutions of equations. Here we shall quote another theorem on the positivity of components of $\mathcal{S}$ (see Auchmuty (1976)).

Assume the reaction terms obey

($\mathcal{F}3$): $f : \Lambda \times \bar{\Omega} \times \mathbb{P}^m \to \mathbb{R}^m$ is continuous and for each

(μ, x, u) in $\Lambda \times \bar{\Omega} \times \partial \mathbb{P}^m$ one has $f_i(\mu, x, u) \geq 0$

for those i such that $u_i = 0$.

<u>Theorem 5</u>.   Let u_λ be a classical solution of $(5.1)-(5.2)$ with $u_{\lambda i}(x) > 0$ for all x in $\bar{\Omega}$ and $1 \leq i \leq m$.   Suppose f obeys ($\mathcal{F}3$) and $\gamma_i(\mu, x) \geq \delta > 0$ on $\partial \Omega$ for all μ in Λ , $1 \leq i \leq m$, then if $\mathcal{S}_k$ is a component of $\mathcal{S}$ which contains (λ, u_λ) and if $(\mu, u_\mu) \in \mathcal{S}_k$ then $u_{\mu i}(x) > 0$ for all x in $\bar{\Omega}$, $1 \leq i \leq m$.

Essentially this theorem says that for these systems, any solution which may be connected to a known positive solution is itself positive. This theorem applies to a number of the models described in Section 4.

Suppose now that one may take $\gamma_i(\lambda,x) = 0$ in (6.2) and let $G_i(x,y)$ be the Green's function for the linear elliptic operator

$$L_i v(x) = -\sum_{j=1}^{n} \frac{\partial}{\partial x_j} (D_i(x)\frac{\partial v}{\partial x_j})$$

subject to $\qquad \alpha_i(x)\frac{\partial v}{\partial \nu} + \beta_i(x)v = 0$.

Then the system of equations (5.1)-(5.2) may be written as a system of nonlinear integral equations

$$v_i(x) - \int_{\Omega} G_i(x,y) f_i(\mu,y,v(y))dy = 0 \quad 1 \le i \le m. \quad (5.6)$$

This may also be written as a fixed point problem

$$v = GN(\mu,v). \quad (5.7)$$

Here $N : \Lambda \times C(\bar{\Omega};\mathbb{P}^m) \to C(\bar{\Omega};\mathbb{R}^m)$ is defined by

$$N(\mu,v)(x) = (f_1(\mu,x,v(x)),\ldots,f_m(\mu,x,v(x)))$$

and $G : C(\bar{\Omega};\mathbb{R}^m) \to C(\bar{\Omega};\mathbb{R}^m)$ is a compact operator with

$$Gv(x) = (G_1 v(x),\ldots,G_m v(x))$$

and $\qquad G_i v(x) = \int_{\Omega} G_i(x,y)v_i(y)dy, \qquad 1 \le i \le m.$

Provided f obeys (𝔉1), N will be a continuous map, and $GN(\mu,\cdot)$ will be a compact map for each μ. This implies that for each value of μ, the solutions of (5.7) form a closed, locally compact set $\mathbf{g}_\mu$ in $C(\bar{\Omega};\mathbb{P}^m)$. Moreover, equation (5.7) is of the form to which one can apply Leray-Schauder degree theory (see Nirenberg (1974)) to obtain information on the existence, and dependence on u , of solutions.

Restricting attention to the case where Λ is an open subset of $\mathbb{R}^1$, we can obtain stronger results if f is also assumed to obey a differentiability condition. For all undefined terms in the next section see Nirenberg (1974). Introduce

(34): $N : \Lambda \times C(\bar{\Omega}; \mathbb{P}^m) \to C(\bar{\Omega}; \mathbb{R}^m)$ is continuously

Fréchet differentiable at each interior point

(u,v) of $\Lambda \times C(\bar{\Omega}; \mathbb{P}^m)$.

This condition will hold if f is uniformly, twice continuously differentiable in u and v on $\bar{\Omega}$. A solution (u_0,v) of (5.7) is said to be a singular point of (5.7) if there is a non-trivial solution w of

$$w = GLw \qquad (5.8)$$

where G as before and $L : C(\bar{\Omega}; \mathbb{R}^m) \to C(\bar{\Omega}; \mathbb{R}^m)$ is the Fréchet derivative of N with respect to u at (u_0,v) . Thus L is linear and

$$Lw(x) = (L_1 w(x),\ldots,L_m w(x))$$

with $\qquad L_i w(x) = \sum_{j=1}^m \frac{\partial f_i}{\partial v_j} (u_0,x,v(x)) w_j(x)$

From the implicit function theorem on Banach spaces, one has that if (u_0,v) is not a singular point of (5.7) then there is a unique curve of solutions, defined for u in an open neighborhood of u_0 , which passes through (u_0,v) .

78   J. F. G. AUCHMUTY

When (μ_0,v) is a singular point, one must make a
careful analysis to describe the solution set near (μ_0,v) .
This solution might, amongst other possibilities, be an
isolated solution, an end-point or a turning point of a
curve of solutions or a bifurcation point for the system.
In this last case, one might have a finite number of curves
or surfaces of solutions passing through (μ_0,v) .

A solution (μ,v) of (5.7) is said to be isolated if
there exists a neighborhood U of (μ,v) with the only
solution of (5.7) in U being (μ,v) .

A solution (μ_0,v) is an end-point of a curve of
solutions of (5.7) if for all sufficiently small $r>0$,
there is a unique solution of (5.7) on the sphere of radius
r and center (μ_0,v) in $\Lambda \times C(\bar{\Omega};\mathbb{P}^m)$.   It will be a
turning-point of the curve if, for   r   small enough, the
solutions (μ,u) of (5.7) contained in the ball of radius
r and center (μ_0,v) in $\Lambda \times C(\bar{\Omega};\mathbb{P}^m)$ obey $\mu \le \mu_0$ (or
$\mu \ge \mu_0$) .

Let $C = \{(\mu(\epsilon),v(\epsilon)) : |\epsilon| < \epsilon_0\}$ be a continuous curve
of solutions of (5.7) with $\mu(0)=\mu_0$, $v(0) = v_0$.   Then
(μ_0,v_0) is a bifurcation point of (5.7) if in any open
neighborhood U of (μ_0,v_0) there is a solution of (5.7)
which does not lie on C .

Using these definitions and the methods of bifurcation
theory, one can analyze the dependence of the steady-state

solutions of reaction-diffusion equations on various
parameters. This has been done for a number of particular
model systems. The trimolecular scheme has been analyzed
in Auchmuty and Nicolis (1975) and (1976) and by
Herschkowitz-Kaufman (1975). Any further unreferenced
results on the trimolecular scheme come from these papers.
Other work on bifurcation analysis of chemical systems
includes Cohen (1973), Fife (1974), Keener and Keller (1973)
and Poore (1973).

One also needs to introduce various definitions of
stability. A steady-state solution v of (3.1)-(3.3) is,
loosely speaking, a stable solution of the equations if for
any initial condition u_0 which is sufficiently close to
v , in some topology, the corresponding solution $u(x,t)$
of (3.1)-(3.3) is defined for all time and obeys

$$\lim_{t \to \infty} u(x,t) = v(x)$$

in some topology.

When (μ, v_μ) is a classical solution of (5.1)-(5.2),
the linear stability problem for v_μ is to find the eigenvalues λ and eigenfunctions w of the system

$$\sum_{j=1}^{n} \frac{\partial}{\partial x_j} (D_i(x) \frac{\partial w_i}{\partial x_j}) + \sum_{k=1}^{m} F_{ik}(x)w_k = \lambda w_i \quad \text{in } \Omega, \quad 1 \le i \le m$$
(5.9)

with $\qquad \alpha_i(x) \frac{\partial w_i}{\partial \nu} + \beta_i(x)w_i = 0 \quad \text{on } \partial\Omega, \quad 1 \le i \le m.$ (5.10)

Here D_i, α_i and β_i are as before and

$$F_{ik}(x) = \frac{\partial f_i}{\partial v_k} (\mu, x, v_\mu(x)) \ .$$

From elliptic spectral theory (see Georgakis and Sani
(1974)), one may show that the spectrum of this operator
consists only of isolated eigenvalues of finite multiplicity
and the set of all eigenvalues is unbounded.

The solution v_μ is said to be linearly stable if for
any eigenvalue λ of (5.9)-(5.10), the real part of λ ,
$\mathrm{Re}\,\lambda$, is negative. It is marginally stable if $\mathrm{Re}\,\lambda \leq 0$
for any eigenvalue λ . It is linearly unstable if there
is an eigenvalue with $\mathrm{Re}\,\lambda > 0$.

The precise relationship between linear stability and
instability and the actual stability or instability of
steady-state solutions of the nonlinear system (3.1)-(3.3)
is quite technical, see Henry (1976). Henceforth when
stability is mentioned, linear stability should be
understood.

It is worth noting that (μ, v_μ) is a singular point
of (5.6) if and only if 0 is an eigenvalue of the linear
stability equations (5.9)-(5.10).

Suppose λ is a complex eigenvalue of (5.9)-(5.10),
then since the coefficients are real, $\bar{\lambda}$, the complex
conjugate of λ , will also be an eigenvalue. If one takes
a continuous one parameter family $(\mu(\tau), v(\tau))$ of solutions
of (5.1)-(5.2), then the eigenvalues of (5.9)-(5.10) will

also depend continuously on τ (see Kato (1966), Section IV 3.5). As τ varies over an interval, either

(i)   the stability properties of $(\mu(\tau),v(\tau))$ do not change

OR   (ii)   there is a critical value τ_0 of τ such that $v(\tau_0)$ is a marginally stable solution.

The stability is said to change at τ_0 , if $(\tau,u(\tau))$ is stable for τ on one side of τ_0 and unstable on the other side.

To illustrate the application of these concepts, consider the trimolecular scheme. Take B to be the parameter, and assume $A(x) \equiv A_0$ for all x in Ω . Then

$$X(x) \equiv A_0 \qquad Y(x) = BA_0^{-1} \qquad x \in \Omega \qquad (5.11)$$

is a steady state solution of (4.2) subject either to (4.4) or (4.5). Moreover it is the unique steady-state solution of (4.6).

The solution is also the unique steady state solution and is a stable solution of (4.2), for B small enough and positive. For the rest of this section we shall only treat the system for boundary conditions (4.4), but a similar (and in some respects simpler) analysis holds when (4.5) is imposed. Also we shall omit the subscript on the A's. Let

$$B_n = 1 + D_1 D_2^{-1} A^2 + \lambda_n (D_1 + D_2)$$

$$\beta_n = 1 + D_1 D_2^{-1} A^2 + D_2^{-1} \lambda_n^{-1} A^2 + D_1 \lambda_n$$

and $B_H = \min_{n \geq 1} B_n = B_1$, $B_c = \min_{n \geq 1} \beta_n$, $B^* = \min(B_1, B_c)$.

Here λ_n is the n-th eigenvalue of $-\Delta$ on Ω, subject to the boundary condition $u = 0$ on $\partial\Omega$, ordered in the usual way; $\lambda_m \leq \lambda_n$ when $m < n$.

The solution (5.11) is linearly stable provided $B < B^*$. When $B = B^*$ it is marginally stable and when $B > B^*$ it is unstable and the system tends, asymptotically in time, to a state which is either temporally or spatially inhomogeneous (i.e. to a "dissipative structure").

Let n_c be the value of n for which β_n is minimized so $B_c = \beta_{n_c}$. Assume n_c is unique and the corresponding eigenvalue λ_{n_c} is simple. Then at B_c , a new branch of steady state solutions bifurcates from the basic solution (5.11). This new branch of solutions is defined only for $B \geq B_c$ in a small neighborhood of the basic solution if n_c is even and it is defined for all B in a neighborhood of B_c when n is odd, $\Omega = [0,1]$.

Similar results hold at any other value of β_n which corresponds to a simple eigenvalue λ_n . Thus the component of $\mathfrak{s}$ containing $\mathfrak{s}_0$ also contains many other curves which intersect $\mathfrak{s}_0$ at the values $B = \beta_n$, $1 \leq n < \infty$. Near $\mathfrak{s}_0$ the solution diagram may be represented

figuratively by the following

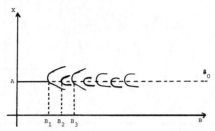

Figure 1. Schematic bifurcation diagram for
trimolecular scheme. The curves of solutions
crossing the basic solution $X \equiv A$ do not
end but their behaviour far from the basic
solution is not known.

For this system there is an analogue of theorem 5

which gives that any solution lying in the component of $\mathfrak{s}$

containing $\mathfrak{s}_0$ is non-negative. This result implies that

any solution which is obtained by bifurcation (and repeated,

or secondary, bifurcation) from the basic solution is

nonnegative and consequently chemically realistic. An a

priori upper bound on the non-negative steady-state

solutions of the equations has been derived. All of these

facts together with a theorem due to Rabinowitz (1971)

implies that each curve C_ℓ of solutions which crosses

$\mathfrak{s}_0$ at $B = \beta_\ell$ either is defined for all $B \geq \beta_\ell$ or else

it returns to the basic solution at some other value β_k

of B . At present, it is not known which of these two

alternatives holds in any particular case.

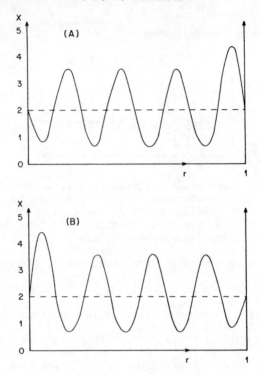

<u>Figure 2</u>.   Examples of two different stable steady-
                state solutions for the trimolecular scheme
                with the same parameters. Here $\Omega = [0,1]$,
                $A = 2$, $B = 4.6$, $D_1 = 1.6 \times 10^{-3}$, $D_2 = 8 \times 10^{-3}$
                and Dirichlet data is used. The solution
                $X \equiv 2$ is unstable (from Herschkowitz-
                Kaufman (1975))

    Mahar and Matkowsky (1976) have recently indicated

that secondary bifurcation occurs in this system, namely,

another curve of solutions can intersect one of the

bifurcated curves of solutions. See also Keener (1976).

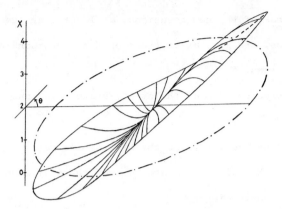

Figure 3. A steady state solution for the trimolecular
scheme on a circle of radius 0.1 with
$A = 2$, $B = 4.6$, $D_1 = 3.25 \times 10^{-3}$, $D_2 = 1.62 \times 10^{-2}$
and no flux boundary conditions.

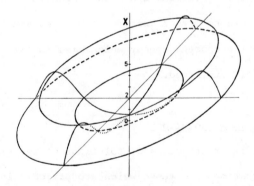

Figure 4. A steady state solution for the trimolecular
scheme on a circle of radius 0.2 with
$A = 2$, $B = 4.6$, $D_1 = 1.6 \times 10^{-3}$, $D_2 = 8 \times 10^{-3}$
and fixed boundary conditions (from Erneux
and Herschkowitz-Kaufman (1975)).

Near the bifurcation points, one may obtain approximate
expressions for the new solutions of the equations. The
leading new term in these expressions is proportional to
the corresponding eigenfunction of the Laplacian. Thus

86 J. F. G. AUCHMUTY

the patterns of the new solutions, at least near the bifurcation, are similar to those of the corresponding eigenfunctions.

The precise form of the steady-state solutions depends on

(i) the geometry and the size of the region Ω,

(ii) the values D_1 and D_2 of the diffusion coefficients

and (iii) the values of B and A and one rate constant k (which was put equal to 1).

Some steady-state solutions are illustrated in figures 2-4.

When A is no longer assumed to be constant in Ω, some different phenomena appear which have not, so far, been completely explained.

VI. Time-periodic Solutions

Another important class of solutions of reaction-diffusion equations are those which are periodic in time. They are of great interest and importance as models for biological clocks and various oscillatory phenomena.

A solution of (3.1) is said to be time-periodic if there is a least $T(>0)$ such that

$$u(x,t+T) = u(x,t) \quad \text{for all } x \text{ in } \Omega, \, t \geq 0 . \quad (6.1)$$

Again it is easiest to analyze these solutions if one introduces a parameter μ taking values in an interval

$\Lambda \subseteq \mathbb{R}^1$. Consider the system of equations

$$\frac{\partial u_i}{\partial t} = \sum_{j=1}^{n} \frac{\partial}{\partial x_j} (D_i(x)\frac{\partial u_i}{\partial x_j}) + f_i(\mu,x,u) \text{ on } \Omega \times (0,T), \qquad (6.2)$$
$$1 \le i \le m.$$

subject to the boundary conditions (5.2), with f_i obeying
the same conditions as in the last section.

When $f_i(\mu,x,u) = g_i(\mu,u)$ is independent of x for
all i then any time-periodic solution $w(t)$ of the
ordinary differential equation

$$\frac{dw_i}{dt} = g_i(\mu,w) \qquad 1 \le i \le m \qquad (6.3)$$

defines a spatially-homogeneous time-periodic solution of
(6.2) by

$$u_i(x,t) \equiv w_i(t) \qquad 1 \le i \le m, x \in \bar{\Omega}. \qquad (6.4)$$

It is a solution of the boundary value problem when
Neumann conditions hold.

The usual method for proving that (6.2) has a time-
periodic solution is to show that as μ varies, the system
(6.2) undergoes a Hopf bifurcation. For introductions to
the theory of Hopf bifurcations for partial differential
equations see Iooss (1973), Ize (1976), Marsden and
McCracken (1976) or Sattinger (1972).

Suppose that for $\mu_0 - \epsilon < \mu < \mu_0 + \epsilon$, there is a
continuous, smooth curve of steady-state solutions $u(\mu)$ of
(5.1)-(5.2). Suppose that $\lambda(\mu)$ is an eigenvalue of the
linear stability equations (5.9)-(5.10) for $u(\mu)$, with

88   J. F. G. AUCHMUTY

$\lambda(\mu)$ being a continuous function of μ near μ_0 and
$\lambda(\mu_0) = i\alpha$, $(\alpha \neq 0)$. Then $\bar{\lambda}(\mu_0) = -i\alpha$ is also an eigenvalue of the linear stability equation. Assume there are no
other eigenvalues of the linear stability equation with
zero real part when $\mu = \mu_0$ and that a transversality
condition holds then, essentially, there is a family of
periodic solutions $v(\mu,x,t)$ of period $T(\mu)$ with $T(\mu)$
converging to $2\pi/\alpha$ and the amplitude of the oscillation
going to zero as μ goes to μ_0 . This situation where
a curve of time-periodic oscillations bifurcates from a
curve of steady-state solutions is called a Hopf bifurcation.
For detailed, careful statements of conditions for a Hopf
bifurcation and approximate expressions for the time-periodic
solutions near the bifurcation point see the above references.

Hopf bifurcations occur in many reaction-diffusion
systems. It can occur in the trimolecular scheme at values
$B_1, B_2, \ldots, B_k$, where k depends on the magnitude of D_1, D_2
and the size of Ω , and is finite if $D_1 \neq D_2$. It is also
possible to find ranges of parameters for the Dirichlet
problem (4.2)-(4.4) for which there is no Hopf bifurcation.
Near the bifurcation point, the time-periodic solutions
generally have a leading new term which is proportional to
the product of $\cos \alpha t$ and the eigenfunction of the linear
stability equation corresponding to the eigenvalue $i\alpha$
(see Auchmuty and Nicolis (1976) for expressions for the

trimolecular scheme). Figures 5 shows one of these time-periodic solutions for the trimolecular scheme.

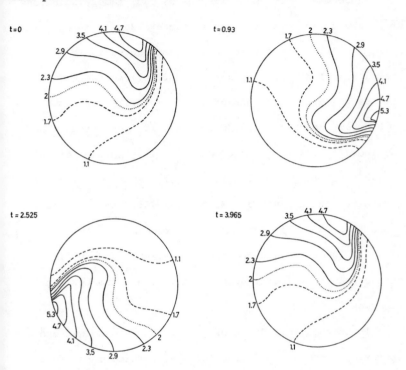

Figure 5. A rotating time-periodic solution of the trimolecular scheme inside a circle with no flux boundary conditions. Here $R = 0.5861$, $A = 2$, $B = 5.8$, $D_1 = 8 \times 10^{-3}$, $D_2 = 4 \times 10^{-3}$. Curves of equal concentration of X are represented by full, dotted or broken lines, respectively, when the concentration is greater than, equal to or less than the trivial solution $X \equiv 2$. (from Erneux and Herschkowitz-Kaufman (1977))

The problem of finding the stability of these time-periodic solutions is treated in most of the above references on the Hopf bifurcation.

There are other special types of oscillatory solutions which are of importance. For instance assume the region $\Omega = \Omega_1 \times [0, 2\pi)$ is periodic in at least one variable. Examples include circles, annuli, tori or spheres. Then one has to solve equations of the form (3.1)-(3.2) subject to boundary conditions and/or periodicity conditions. One may look for solutions which are wave-like in the periodic variable. Turing (1952) first analyzed such equations on a circle. In Auchmuty and Nicolis (1976) Section 6, it is shown that, on a circle, there are solutions of the form $u(\theta - vt)$ where θ is the angular variable and v is a characteristic velocity determined by the parameters in the equation.

In another vein Greenberg (1975) has shown that one can approximate certain spiral solutions of reaction-diffusion equations. Other results on the patterns and rhythms possible in reaction-diffusion equations have been obtained by Gmitro and Scriven (1965), Othmer and Scriven (1969) and by Sattinger (1976).

VII. Conclusion

The analysis in this paper has mostly been concerned with the positivity and the bifurcation-theoretic properties of reaction-diffusion systems. We have seen that appropriate

generalizations of criteria for positivity of chemical
kinetic equations carry over to give positivity for reaction-
diffusion systems. Diffusion effects, however, make the
bifurcation properties of reaction-diffusion equations much
more complicated than the same systems without diffusion.
One may obtain a great variety of patterns and oscillations
or waves as asymptotic, in time, solutions and the particular
behaviour observed depends not only on the chemical scheme
but also on the diffusivity, the geometry and size of the
region and the boundary conditions.

Unfortunately at present, we are in no position to give
a complete classification of the types of patterns which
may be obtained in reaction-diffusion systems, or to
describe general conditions under which particular types of
behaviour occur.

Nevertheless, for morphogenesis and some other
biological problems, these analyses have already yielded
a number of implications.

There is now no question that Turing was correct in
believing that coupled reaction and diffusion in a system
can yield intricate chemical patterns and/or oscillations or
waves. Moreover this can occur autonomously. There is no
need to impose external patterns or a forcing function to
get a system to display similar patterns or oscillations.
In particular, one need not impose a morphogenetic gradient

for a system to evolve to a spatially inhomogeneous solution.
(see Babloyantz and Hiernaux (1976)).

   This is an example of symmetry-breaking. The fact that
equations have certain symmetries does not necessarily
imply that all solutions of the equations have the same
symmetries, or that the solutions with these symmetries are
stable. This phenomenon is associated with nonlinear
equations and has often been commented on in hydrodynamics
(see Birkhoff (1950)). The coefficients in a system of
nonlinear reaction-diffusion equations may be constant in
space and time but the system may evolve to states which
are not.

   It is also instructive to compare reaction-diffusion
models with Thom's (1972) theories of morphogenesis.

   Firstly reaction-diffusion models are models in the
traditional scientific sense. One posits specific physical
and chemical processes and, using the laws governing these
processes, one derives equations which describe the evolution
as well as the asymptotic behaviour of the systems. Analysis
of the equations yields specific predictions which may be
compared with experiments in a quantitative, as well as a
qualitative, manner.

   Having established a specific model, one can study the
stability of the solutions to various types of perturbations.
One does not need a blanket requirement of structural

stability. It is too general. A reasonable requirement is that models of biological systems be consistent with the physics and chemistry of the systems. For reaction-diffusion equations one need not study their stability to arbitrary perturbations but only to those which are consistent with the laws of chemical kinetics and the observed properties of diffusion.

Also the solutions of reaction-diffusion equations may depend on the parameters involved in many other ways besides those enumerated by the elementary catastrophes. For example, as parameters are changed, stable steady state solutions may give way to time-periodic oscillations. In contrast, gradient systems can never have any time-periodic solutions.

Finally, I hope that this summary of some of the recent work on reaction-diffusion equations has indicated some of the richness of structure and the wealth of phenomena that may be found in these systems.

The author would like to thank G. Nicolis and H. G. Othmer for valuable comments during the preparation of this paper and M. Herschkowitz-Kaufman and T. Erneux for permission to reproduce figures. This research was partially supported by NSF grant MCS76-07273.

DEPARTMENT OF MATHEMATICS
INDIANA UNIVERSITY
BLOOMINGTON, INDIANA   47401

BIBLIOGRAPHY

Amundson, N. R. (1974). Nonlinear Problems in Chemical
    Reactor Theory, in SIAM-AMS Proceedings, Vol. 8,
    pp. 59-84, Providence.

Aris, R. (1965). Prolegomena to the rational analysis of
    systems of chemical reactions, Arch. Rat. Mech. Anal.
    19, pp. 81-99.

_____, (1968). Prolegomena to the rational analysis of
    systems of chemical reactions II. Arch. Rat. Mech.
    Anal. 27, pp. 356-364.

_____, (1975). The Mathematical Theory of Diffusion and
    Reaction in Permeable Catalysts, 2 Vols., Oxford U.P.,
    Oxford.

Aronson, D. G. and H. F. Weinberger (1975). Nonlinear
    diffusion in population genetics, combustion and nerve
    propagation. Lecture Notes in Math., Vol. 446,
    Springer Verlag, Heidelberg.

Auchmuty, J. F. G. (1973). Lyapunov methods and equations
    of parabolic type, Lecture Notes in Math. Vol. 322,
    pp. 1-14, Springer Verlag, Heidelberg.

_____, (1977). Positivity for Elliptic and Parabolic
    Systems. Proc. Roy. Soc. Edinburgh 79A, pp. 183-191.

Auchmuty, J. F. G. and G. Nicolis (1975). Bifurcation
    Analysis of Nonlinear Reaction-Diffusion Equations - I.
    Evolution Equations and the Steady State Solutions,
    Bull. Math. Biol. 37, pp. 323-365.

_____, (1976). Bifurcation Analysis of Nonlinear Reaction-
    Diffusion III Chemical Oscillations. Bull. Math. Biol.
    38, pp. 325-350.

Babloyantz, A. and J. Hiernaux (1976). Models for Cell
    Differentiation and Generation of Polarity in Diffusion
    Governed Morphogenetic Fields (preprint).

Balslev, I. and Degn, H. (1975). Spatial Instability in
    Simple Reaction Systems, J. Theor. Biology 49, pp. 173-
    177.

Birkhoff, G. (1950). Hydrodynamics, A Study in Logic, Fact,
    and Similitude, Princeton.

Caplan, S. R., A. Naparstek and N. J. Zabusky (1973).
    Chemical Oscillations in a Membrane, Nature 245,
    pp. 364-366.

Chueh, K. N., C. C. Conley and J. A. Smoller (1977).
    Positively Invariant Regions for Systems of Nonlinear
    Diffusion Equations, Ind. J. Math. 26, pp. 373-392.

Cohen, D. S. (1973). Multiple Solutions of Nonlinear Partial
    Differential Equations, Lecture Notes in Math. Vol. 322
    pp. 15-77, Springer Verlag, Heidelberg.

Conway, E., D. Hoff and J. Smoller (1977). Large Time
    Behaviour of Solutions of Systems of Nonlinear Reaction
    Diffusion Systems (preprint).

Conway, E. and J. Smoller (1977). Diffusion and the
    Classical Ecological Interactions: Asymptotics
    (preprint).

Crandall, M. G. and P. H. Rabinowitz (1970). Nonlinear
    Sturm Liouville Eigenvalue Problems and Topological
    Degree, J. Math. Mech. 19, pp. 1083-1102.

Dancer, E. N. (1973). Global Solution Branches for Positive
    Mappings, Arch. Rat. Mech. and Anal. 52, pp. 181-192.

Diekmann, O. and N. M. Temme (eds) (1976). Nonlinear
    Diffusion Problems, Math. Centrum, Vol. 28, Amsterdam.

Duvaut, G. and J. L. Lions (1972). Les Inéquations en
    Mécanique et en Physique, Dunod, Paris.

Erneux, T. and Herschkowitz-Kaufman M. (1975). Dissipative
    Structures in Two Dimensions. Biophysical Chemistry,
    3, pp. 345-354.

_____, (1977). Rotating Waves as Asymptotic Solutions
    of a Model Chemical Reaction, J. Chem. Phys. 66 pp. 248.

Erneux, T., J. Hiernaux and G. Nicolis (1977). Turing's
    Theory of Morphogenesis (preprint).

Fife, P. C. (1974). Branching Phenomena in Fluid Dynamics
    and Chemical Reaction Diffusion Theory in "Eigen-
    values of Nonlinear Problems", CIME, Cremonese Rome
    pp. 25-83.

Friedman, A. (1965). Remarks on Nonlinear Parabolic Equations
    Proc. Symp. in Applied Math. Vol. 17, pp. 3-23.

Gavalas, G. R. (1969). Nonlinear Differential Equations of
    Chemically Reacting Systems, Springer Verlag, New York.

Georgakis, C. and R. L. Sani (1974). On the Stability of
    the Steady State in Systems of Coupled Reaction and
    Diffusion, Arch. Rat. Mech. Anal. 52, pp. 266-296.

Gierer, A. and H. Meinhardt (1972). A Theory of Biological
    Pattern Formation, Kybernetik 12, pp. 30-39.

_____, (1974). Biological Pattern Formation Involving
    Lateral Inhibition, Some Math Questions in Biology
    VI, A.M.S., Providence.

Glansdorff, P. and I. Prigogine (1971). Thermodynamic
    Theory of Structure, Stability and Fluctuations, Wiley-
    Interscience, London.

Gmitro, J. I. and L. E. Scriven (1965). A Physicochemical
    Basis for Pattern and Rhythm in Intracellular Transport
    ed. K. B. Warren, pp. 221-255, Academic Press, New York.

Goldbeter, A. (1973). Patterns of Spatio-Temporal
    Organization in an Allosteric Enzyme Model, Proc. Nat.
    Acad. Sci. U.S.A. 70, pp. 3255-3259.

Greenberg, J. M. (1976). Periodic Solutions to Reaction-
    Diffusion Systems, SIAM J. Appl. Math. 30, pp. 199-205.

Henry, D. (1976). Geometric theory of semilinear parabolic
    equations, Lecture Notes, U. of Kentucky.

Herschkowitz-Kaufman M. (1975). Bifurcation Analysis of
    Nonlinear Reaction-Diffusion Equations II. Steady
    State Solutions and Comparison with Numerical
    Simulations, Bull. Math. Biol. 37, pp. 589-636.

_____, and G. Nicolis (1972). Localized Spatial Structures
    and Nonlinear Chemical Waves in Dissipative Systems.
    J. Chem. Phys. 56, pp. 1890-1895.

Horn, F. and R. Jackson (1972). General Mass Action Kinetics.
    Arch. Rat. Mech. and Anal. 47, pp. 81-116.

Howard, L. N. and N. Kopell (1973). Plane Wave Solutions to
    Reaction-Diffusion Equations, Studies in Appl. Math.
    52, pp. 291-328.

Iooss, G. (1973).   Bifurcation et Stabilité, Lecture Notes
    No. 31, Universite Paris XI.

Ize, G. (1976).   Bifurcation Theory for Fredholm Operators,
    Memoirs A.M.S. No. 174 Providence.

Kato, T. (1966).   Perturbation Theory for Linear Operators,
    Springer Verlag, New York.

Keener, J. P. (1976).   Secondary Bifurcation in Nonlinear
    Diffusion-Reaction Equations, Studies in Applied Math.,
    55, pp. 187-211.

Keener, J. P. and H. B. Keller (1973).   Perturbed Bifurcation
    Theory, Arch. Rat. Mech. Anal. 50, p. 159-175 .

Keller, E. F. and L. A. Segel (1971).   Model for Chemotaxis,
    J. Theor. Biol. 30, pp. 225-234.

Krambeck, F. J. (1970).   The mathematical structure of
    chemical kinetics in homogeneous single-phase systems.
    Arch. Rat. Mech. Anal. 38, pp. 317-347.

Lefever, R. and I. Prigogine (1968).   Symmetry Breaking
    Instabilities in Dissipative Systems II, J. Chem. Phys.
    48, pp. 1695-1700.

Levin, S. A. (1976).   Spatial Patterning and the Structure
    of Ecological Communities, Lectures on Math in the
    Life Sciences Vol. 8, A.M.S. Providence.

Lions, J. L. (1969).   Quelques méthodes de résolution des
    problèmes aux limites non linéaires, Dunod, Paris.

Mahar, T. J. and B. J. Matkowsky (1976).   A model biochemical
    reaction exhibiting secondary bifurcation. SIAM J.
    Appl. Math. 32, pp. 394-404.

Martin, R. H. (1973).   Differential Equations on Closed
    Subsets of Banach Space, Trans. A.M.S. 179, pp. 399-
    414.

Marsden, J. E. and M. McCracken (1976).   The Hopf Bifurcation
    and its Applications, Springer Verlag, New York.

Meurant, G. and J. C. Saut (1977).   Bifurcation and Stability
    in a Chemical System, J. Math. Anal. and Appns. (to
    appear).

Nagumo, N. (1942). Uber die Lage der Integralkurven gewohnlicher Differentialgleichungen, Proc. Phys. Math. Soc. Japan 24, pp. 551-559.

Nicolis, G. and I. Prigogine (1977). Self-organization in Nonequilibrium Systems, Wiley-Interscience, New York.

Nirenberg, L. (1974). Topics in Nonlinear Functional Analysis Lecture Notes, Courant Institute of Math. Sciences, N. Y. U., New York.

Oster, G. F. and A. S. Perelson (1974). Chemical Reaction Dynamics I. Arch. Rat. Mech. Anal. 55, pp. 230-274.

Othmer, H. G. (1976a). Non-uniqueness of Equilibria in Closed Reacting Systems, Chem. Eng. Sci. 31, pp. 993-1003.

Othmer, H. G. (1976b). Current problems in Pattern Formation, Lectures on Math in the Life Sciences, Vol. 9, A.M.S., Providence.

Othmer, H. G. and L. E. Scriven (1969). Interactions of Reaction and Diffusion in Open Systems. Ind. and Eng. Chem. Fund. 8, pp. 302-313.

Poore, A. B. (1973). Multiplicity, Stability and Bifurcation of Periodic Solutions in problems arising from chemical reactor theory, Arch. Rat. Mech. Anal. 52, pp. 358.

Rabinowitz, P. H. (1971). Some Global Results for Nonlinear Eigenvalue Problems, J. Funct. Analysis 7, pp. 487-513.

Rashevsky, N. (1960). Mathematical Biophysics 2 vols., 3rd revised edition. U. of Chicago Press, Chicago.

Redheffer, R. M. and W. Walter (1975). Flow Invariant sets and differential inequalities in normed spaces. Applicable Analysis 5, pp. 149-161.

Rubinow, S. (1975). Introduction to Mathematical Biology. John Wiley, New York.

Sattinger, D. H. (1972). Topics in Stability and Bifurcation Theory, Lecture Notes in Math. Vol. 309, Springer Verlag, Heidelberg.

_____, (1976). Group Representation Theory, Bifurcation Theory and Pattern Formation, U. of Minnesota, preprint.

Scribner, A., L. A. Segel and E. H. Rogers (1974). A Numerical
    Study of the Formation and Propagation of Traveling
    Bonds of Chemotactic Bacteria. J. Theor. Biol. 46,
    pp. 189-219.

Segel, L. A. and J. Jackson (1972). Dissipative structure;
    an explanation and an ecological example. J. Theor. Biol.
    37, pp. 545-559.

Smale, S. (1974). Mathematical Model of two cells via
    Turing's Equation, Some Math. Questions in Biology
    V, A.M.S. Providence.

Thom, R. (1972). Stabilité Structurelle et Morphogénèse,
    W. A. Benjamin Inc., Reading.

Turing, A. (1952). The Chemical Basis of Morphogenesis,
    Phil. Trans. Roy. Soc. B237, pp. 37-72.

Turner, R. E. L. (1975). Transversality and Cone Maps,
    Arch. Rat. Mech. and Anal. 58, pp. 151-179.

Tyson, J. J. (1976). The Belousov-Zhabotinskii Reaction.
    Lecture Notes in Biomathematics Vol. 10, Springer
    Verlag, Berlin.

Wallwork, D. and A. S. Perelson (1976). Restrictions on
    Chemical Kinetic Models, J. Chem. Phys. 65, pp. 284-
    292.

Weinberger, H. F. (1975). Invariant Sets for Weakly
    Coupled Parabolic and Elliptic Systems. Rend. di. Mat.
    Ser. VI 8, pp. 295-310.

Williams, S. A. and P. L. Chow (1977). Nonlinear Reaction-
    Diffusion Models for Interacting Populations (preprint).

Winfree, A. T. (1974a). Rotating Solutions to Reaction
    Diffusion Equations in Simply-Connected Media.
    SIAM-AMS Proceedings Vol. 8, Providence, pp. 13-31.

Winfree, A. T. (1974b). Two kinds of wave in an oscillating
    chemical solution, Trans. Faraday Symposia of the
    Chemical Soc. 9, pp. 38-46.

Lectures on Mathematics in the Life Sciences
Volume 10, 1978

REPRESENTING VISUAL INFORMATION

D. Marr

SUMMARY: Vision is the construction of efficient
symbolic descriptions from images of the world.   An important
aspect of vision is the choice of representations for the
different kinds of information in a visual scene.   In the
early stages of the analysis of an image, the representations
used depend more on what it is possible to compute from an
image than on what is ultimately desirable, but later
representations can be more sensitive to the specific needs of
recognition.   This essay surveys recent work in vision at
M. I. T. from a perspective in which the representational
problems assume a primary importance.   An overall framework is
suggested for visual information processing, in which the
analysis proceeds through three representations; (1) the
primal sketch, which makes explicit the intensity changes and
local two-dimensional geometry of an image, (2) the $2\frac{1}{2}$-D
sketch, which is a viewer-centered representation of the
depth, orientation and discontinuities of the visible
surfaces, and (3) the 3-D model representation, which allows
an object-centered description of the three-dimensional
structure and organization of a viewed shape.   Recent results
concerning processes for constructing and maintaining these
representations are summarized and discussed.

Contents

0: Introduction

0.1: Understanding information processing tasks

Vision is an information processing task, and like any
other, it needs understanding at two levels. The first, which
I call the computational theory of an information processing
task, is concerned with what is being computed and why; and
the second level, that at which particular algorithms are
designed, with how the computation is to be carried out (Marr
& Poggio 1977a). For example, the theory of the Fourier
transform is a level 1 theory, and is expressed independently
of ways of obtaining it (algorithms like the Fast Fourier
Transform, or the parallel algorithms of coherent optics) that
lie at level 2. Chomsky calls level 1 theories competence
theories, and level 2 theories performance theories. The
theory of a computation must precede the design of algorithms
for carrying it out, because one cannot seriously contemplate
designing an algorithm or a program until one knows precisely
what it is meant to be doing.

I believe this point is worth emphasizing, because it
is important to be clear about the level at which one is
pursuing one's studies. For example, there has recently been
much interest in so-called cooperative algorithms (Marr &
Poggio 1976) or relaxation labelling (Rosenfeld, Hummel &
Zucker 1976). The attraction of this technique is that it
allows one to write plausible constraints directly into an
algorithm, but one must remember that such techniques amount

to no more than a style of programming, and they lie at the
second of the two levels. They have nothing to do with the
theory of vision, whose business it is to derive the
constraints and characterize the solutions that are consistent
with them.

0.2: Understanding vision

If one accepts in broad terms this statement of what
it means to understand an information processing task, one can
go on to ask about the particular theories that one needs to
understand vision.   Vision can be thought of as a *process*,
that produces from images of the external world a description
that is useful to the viewer and not cluttered by irrelevant
information. These descriptions, in turn, are built or
assembled from many different but fixed representations, each
capturing some aspect of the visual scene. In this article, I
shall try to present a summary of our work on vision at M. I. T.
seen from a perspective in which the representational problems
assume a primary importance. I shall include summaries of our
present ideas as well as of completed work.

The important point about a representation is that it
makes certain information *explicit* (cf the principle of
explicit naming, Marr 1976). For example, at some point in
the analysis of an image, the intensity changes present there
need to be made explicit, so does the geometry -- of the image
and of the viewed shape -- and so do other parameters like
color, motion, position and binocular disparity. To
understand vision thus requires that we first have some idea
of which representations to use, and then we can proceed to
analyze the computational problems that arise in obtaining and
manipulating each representation. Clearly the choice of
representation is crucial in any given instance, for an

inappropriate choice can lead to unwieldy and inefficient
computations. Fortunately, the human visual system offers a
good example of an efficient vision processor, and therefore
provides important clues to the representations that are most
appropriate and likely to yield successful solutions.

     This point of view places the nature of the
representations at the center of attention, but it is
important to remember that the limitations on the processes
that create and use these representations are an important
factor in determining their structure, because one of the
constraints on vision is that the description ultimately
produced be derivable from images. In general, the structure
of a representation is determined at the lower levels mostly
by what it is possible to compute, whereas later on they can
afford to be influenced by what it is desirable to compute for
the purposes of recognition.

1: Early processing problems

1.0: The primal sketch

There are two important kinds of information contained in an intensity array, the intensity changes present there, and the local geometry of the image. The primal sketch (Marr 1976) is a primitive representation that allows this information to be made explicit. Following the clues available from neurophysiology (Hubel & Wiesel 1962), intensity changes are represented by blobs and by oriented elements that specify a position, a contrast, a spatial extent associated with the intensity change, a weak characterization of the type of intensity change involved, and a specification of points at which intensity changes cease (so-called termination points). The representation of local geometry makes explicit two-dimensional geometrical relations between significant items in an image. These include parallel relationships between nearby edges, and the relative positions and orientations of significant places in the image. These significant places are marked by "place-tokens", and they are defined in a variety of ways, by blobs or local patches of different intensity, by small lines, and by the ends of lines or bars. The local geometrical relations between place-tokens are represented by inserting virtual lines that join nearby place-tokens, thus making explicit the existence of a relation between the two tokens, their relative orientation, and the distance between them (Marr 1976 figure 12a).

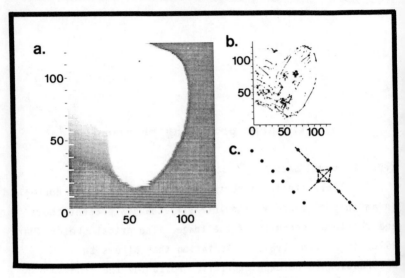

Reprinted from D. Marr, *Early processing of visual information,*
Philosophical Transactions of the Royal Society of London, Ser. B.,
vol. 275, (1976) p. 495 and p. 502.

1.   The primal sketch makes explicit information held in an
intensity array (1a).   There are two kinds, one concerns the
changes in intensity, and this is represented by oriented
edge, line and bar elements, associated with which is a
measure of the contrast and spatial extent of the intensity
change.   The other kind of information is the local two-
dimensional geometry of significant places in the image.   Such
places are marked by "place-tokens", which can be defined in a
variety of ways, and the geometric relations between them are
represented by inserting "virtual lines" between nearby
tokens.   (Marr 1976 figures 7 and 12a).

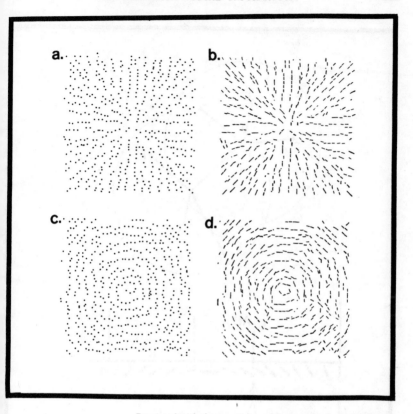

Reprinted with the permission of Springer-Verlag, publishers.

2. 2a and 2c are random-dot interference patterns of the kind described by Glass (1969). 2b and 2d exhibit the results of running the algorithm described in the text and figure 3. The neighborhood radius was such that roughly 8 neighbors were included. (Stevens 1977 figure 5).

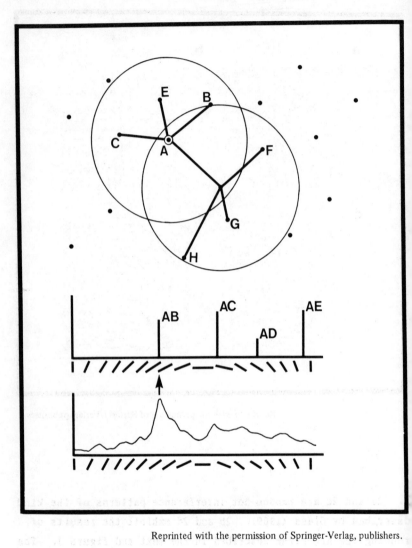

3.   The algorithm for computing locally parallel structure has
three fundamental steps. Place tokens that are defined in the
image are the input to the algorithm, which is applied in
parallel to each one. Since, in the case of the Moiré dot
patterns, each dot contibutes a place token, the first step is
to construct a virtual line from that dot to each neighboring
dot (within some neighborhood centered on the dot). A virtual
line represents the position, separation, and orientation
between a pair of neighboring dots. To favor relatively
nearer neighbors, relatively short virtual lines are
emphasized. The second step is to histogram the orientations
of the virtual lines that were constructed for each of the
neighbors. For example, the neighbor D would contribute
orientations AD, DF, DG, and DH to the histogram. The final
step (after smoothing the histogram) is to determine the
orientation at which the histogram peaks, and to select that
virtual line (AB) closest to that orientation as the solution.
(Stevens 1977 figure 4).

1.1: Random-dot interference patterns

The idea of place-tokens and of this way of representing geometrical relations arose from considering the computational problems that are posed by early visual processing, and one of the questions we have been asking is, can one find any psychophysical evidence that the human visual system makes use of a similar representation? We have recently obtained two results related to this point. Stevens (1977) has examined the perception of random-dot interference patterns (figure 2), constructed by superimposing two copies of a random dot pattern where one copy has undergone some composition of expansion, translation, or rotation transformations (Glass 1969). He found that a simple algorithm suffices to account quantitatively for human performance on these patterns. The algorithm consists of three steps:

(1) Each dot defines a place-token. For example some dots can be replaced by small lines or larger blobs without disrupting the subjective impression of flow.

(2) Virtual lines are inserted between nearby place-tokens, and the neighborhood in which the virtual lines are inserted depends in a predictable way on the density of the dots.

(3) The orientations of the virtual lines attached to all the points in each neighborhood are histogrammed, and locally parallel organization is found by searching for a peak in this histogram. The bucket width that best matches human performance is about 10 degrees.

The details of these steps are set out in figure 3. The interesting features of the algorithm are; (a) It is not iterative. Stevens could find no evidence that human performance rests on a cooperative algorithm, although this type of problem is ideal for that approach. (b) The algorithm

is purely local.   No global-to-local or top-down interactions
are necessary to explain human performance. (c) What the
algorithm finds is locally parallel organization.   In this
case, the organization lies in the virtual lines constructed
between nearby dots, but locally parallel organization among
the real edges and lines in an image also forms an important
part of the structure of an image (Marr 1976).

1.2: Texture discrimination

The second study is one by Schatz (1977) on texture
vision discrimination.   Marr (1976) suggested that such
discriminations could be carried out by first-order
discriminations acting on the description in the primal sketch
(p. 501).   Marr supposed that certain grouping processes were
needed before the discriminations are made in order to account
for the full range of human texture discrimination, but in a
careful examination of the problem, Schatz found that many of
the examples he constructed could be explained by assuming
that the discriminations are made only on real edges or on
virtual lines inserted between neighboring place-tokens.   If
this were generally true, it would stand in elegant relation
to Julesz's (1975) conjecture, that a necessary condition for
the discriminability of two textures is that their dipole
statistics differ.   This condition is known not to be
sufficient, a state of affairs that one can view as implying
that we have access to only a proper subset of all dipole
statistics.   It is possible that this proper subset consists
only of real edges and of the virtual lines that join nearby
place-tokens.

1.3: Discrimination ability

If one accepts that texture discrimination relies upon

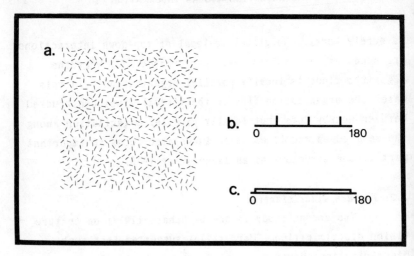

4. The pattern 4a contains two regions, one of whose line segments has the orientation distribution shown in 4b, and the other has the distribution 4c. Surprisingly, three orientations cannot be distinguished from a random orientation distribution. If human texture discrimination is based on first-order discriminations acting on the description held in the primal sketch, the discriminants that can be brought to bear on this information are weak.

first-order discriminations of this type, it is natural to ask
how sensitive are the particular discrimination functions that
we can bring to bear on an image. Riley (1977) has found
evidence that the available functions are extremely coarse.
For example, figure 4 consists of a background in which the
line segments have a random orientation, surrounding a square
containing lines of only three orientations. Surprisingly,
the square cannot be discerned without scrutiny. One
interpretation of this and related findings is, that
discriminations on orientations other than horizontal and
vertical are made on the output of 5 channels, each nearly
binary, and with an angular width of about 35 degrees -- in
other words, only very little information is available about
the distribution of orientations in an image. It appears that
our discrimination ability is as poor or poorer for the other
stimulus dimensions, for example intensity distribution (Riley
1977).

1.4: Light source effects

In another study concerned with what can be extracted
from an image, Ullman (1976a) enquired about the possible
physical basis for the subjective quality of fluorescence,
which is normally associated with the presence of a light
source. He noted that at a light source boundary, the ratio
of intensity to intensity gradient changes sharply, whereas
this is not true at reflectance boundaries unless the surface
orientation changes sharply. He showed that, in the mini-
world of Mondrians, the discriminant to which this leads
predicts human performance satisfactorily.

K. Forbus (in preparation) has extended this work to
the detection of surface luster. Since glossiness is due to
the specular component of a surface reflectivity function, one
can treat the detection of gloss as essentially the detection

of light sources that appear reflected in a surface (see Beck
1974), and this depends ultimately on the ability to detect
light sources. Forbus divided the problem into three
categories; (a) in which the specularity is too small to allow
gradient measurements, (b) in which both intensity and
gradient measurements are available, but the specularity is
local (as it is for a curved surface or a point source), and
(c) in which the surface is planar and the source is extended.
He derived diagnostic criteria for each case.

1.5: Regions from a discriminant

Whenever a region is defined in an image by a
predicate, for example by a difference in texture or
brightness, one faces the problem of delimiting the region
accurately. There are two approaches to designing algorithms
for this problem; one is to use the predicate directly,
deciding whether a given location lies within or without the
region by testing some function of the predicate there. The
second approach is to differentiate the predicate, defining
the region by its boundaries rather than by properties of its
interior.

The difficulties with the problem arise because one is
usually ignorant beforehand of the scale at which significant
predicate signals may be gathered. For example, suppose one
wished to find the boundary between two regions that are
distinguished by different densities of dots. Dot density has
to be measured by selecting a neighborhood size and counting
the number of dots that lie within it. If the neighborhood
size is too large, one may not be able to resolve the regions.
If it is so small as to contain zero, one or two dots, natural
fluctuations may obscure any changes in density.

One solution to this problem is to make the

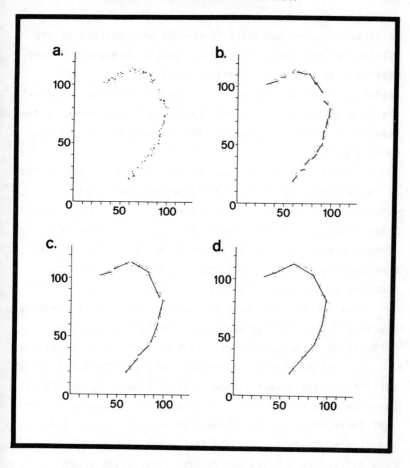

5. Finding a boundary from dot (or place-token) density
changes. Once a rough assignment of boundary points has been
made (5a), local line-fitting (5b) and grouping (5c & d)
techniques can recover a rough specification of the boundary
quite easily.

measurements simultaneously at several neighborhood sizes,
looking for agreement between the results obtained in those
neighborhood sizes that lie just above the size at which
random fluctuations appear. This technique can be applied to
region finding or to boundary finding, and an example of the
results is given in figure 5. The dot density here is not
known *a priori*.

This issue is of considerable techical interest, but
it is important not to lose sight of the underlying
computational problem, which is what kind of boundary is to be
found, and why? The techniques of O'Callaghan (1974) for
example are designed to find boundaries in dot patterns so
accurately that their positions are determined up to the
decision about which dots it passes through. The
justification for this type of study is that humans can assign
boundaries this accurately, but the difficulty lies in
formulating a reasonable definition of what the boundary is.

This problem is a deep one, touching the heart of the
question of what early vision is *for*. I shall return to it
later in this essay, but it is perhaps worth remarking here
that there seems to be a clear need for being able to do early
visual processing roughly and fast as well as more slowly and
accurately, which means having ways of handling rough
descriptions of regions -- ways of characterizing their
approximate extent and shape -- *before* characterizing their
precise boundaries. Figure 6 contains one example of a region
whose rough extent is clear, but whose exact boundary is not.

The motivation for wanting this is that rough
descriptions are very useful during the early stages of
building a shape description for recognition (Marr & Nishihara
1977). For example a man often appears as a roughly vertical
rectangle in an image, and this information is useful because

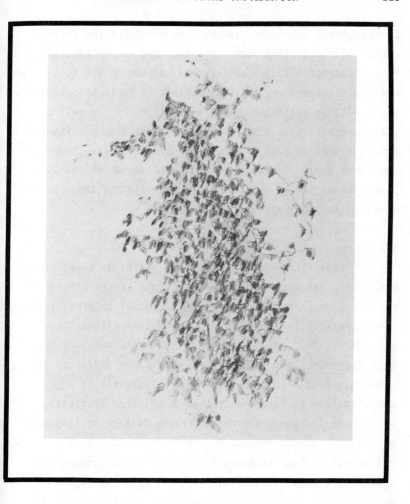

5. An example of a region whose rough boundary is clear, but
whose exact boundary is not. Drawing by K. Prendergast.

it eliminates many other shapes from consideration quite
early. Campbell (1977) has suggested that the extraction of
rough descriptions from an image may depend on the ability to
examine its lower spatial frequencies. Even if this is one of
the available mechanisms it is unlikely to be the only one,
because sparse line drawings can raise the same problems while
having almost no power in their low frequencies. It may be
that some notion of rough grouping applied to low resolution
place-tokens set up by pieces of contour in the image provides
a useful approach to this problem.

1.6: Lightness

Ever since Ernst Mach noticed the bands named after
him, there has been considerable interest in the problem of
computing perceived brightness. Of especial interest is the
recent work of Land & McCann (1971) on the retinex theory (see
also Horn 1974), which is concerned with the quantity they
call lightness; and that of Colas-Baudelaire (1973) on the
computation of perceived brightness. Lightness is an
approximation to reflectance that is obtained by filtering out
slow intensity changes, the underlying idea being that these
are usually due to the illuminant, not to changes in
reflectance. The problem with this idea is of course that
some slow changes in intensity are perceptually important (see
Horn 1977 for an analysis of shape from shading). The linear
filter model of Colas-Baudelaire performs well on images in
which there are no sharp changes in intensity, but the author
found it difficult to extend his model to the more general
case. The recent finding of Gilchrist (1977), that perceived
depth influences perceived brightness, suggests that some
aspects of the problem occur quite late -- in our terms, at
the level of the $2\frac{1}{2}$-D sketch (see below).

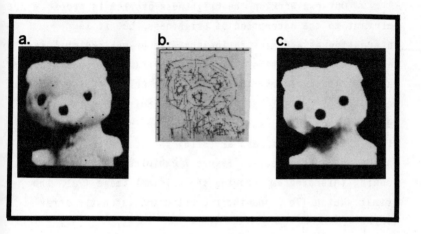

7. An image (7a), the spatial components of its primal sketch (7b), and a reconstruction of the image from the primal sketch (7c). This shows that the our current primal sketch programs lose little of the information in an image.

D. MARR

Our own work on the brightness problem is probably not relevant to the perception of brightness, but it is interesting as a demonstration that the primal sketch loses very little information. Woodham & Marr (unpublished program) have written a program that inverts the primal sketch, so that its output is an intensity array. The basic idea is to scan outwards from edges, assigning a constant brightness to points along the scan lines, and arresting the scan when it encounters another edge. Figure 7 exhibits the results of running this program, showing the original image (7a), the primal sketch (7b), and the reconstructed intensity array (7c).

2: Process-oriented theories

2.0: Introduction

I said earlier that, especially at the earlier stages
of visual information processing, the representations and
processes are determined more by what it is possible to
compute from an image than by what is desirable. Examples are
the problems associated with structure from motion,
stereopsis, texture gradients, and shading.

2.1: Structure from motion

Given a sequence of views of objects in motion, the
human visual system is capable of interpreting the changing
views in terms of the shapes of the viewed objects, and their
motion in three-dimensional space. Even if each successive
view is unrecognizable, the human observer easily perceives
these views in terms of moving objects (Wallach & O'Donnell
1953). To answer the question of how a succession of images
yields an interpretation in terms of three-dimensional
structure in motion, Ullman (1977) divided the problem into
two parts: (1) finding a correspondence between elements in
successive views; and (2) determining the three-dimensional
structures and their motion from the way corresponding
elements move between views.

An important preliminary question about the
correspondence problem concerns the level at which it takes
place. Is it primarily a low-level relation, established

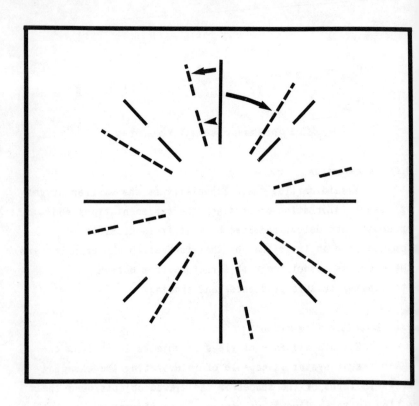

8.   Evidence that the correspondence problem for apparent
motion involves matching operations that act at a low-level.
Frame 1 is shown with full lines and frame 2 with dotted
lines. Instead of seeing a single wheel rotating, when
appropriately timed the wheel splits, the outer and inner
rings rotating one way, and the center rotating the other, as
indicated by the arrows.   This suggests that matching is
carried out on elemental line segments, and is governed
primarily by proximity. (Adapted from Ullman 1977).

between small and simple parts of the scenes and largely
independent of higher-level knowledge and three-dimensional
interpretation? Or do higher level influences, like the
interpretation of the whole of a shape from one frame, play an
important part in determining the correspondence?

Ullman has assembled a considerable amount of evidence
that the former view is correct. For example, figure 8 shows
two successive frames, one denoted with full lines and the
other with dotted lines. If the whole pattern were being
analyzed from one frame, the shape of the wheel extracted, and
used to match the elements in the next frame, the observer
presented with these frames in rapid succession should
perceive them as a whole wheel rotating. Notice however that
the inner and outer parts of the wheel have their closest
neighbors in one direction, whereas the center parts have
theirs in the other; because of this, if the matching were
done early and locally, the observer should see the center
part rotating one way, and the inner and outer rings rotating
the other (as shown with arrows in figure 8). When
appropriately timed, this is in fact what happens.

Another line of evidence is the following. The most
important factor in finding a correspondence between elements
is the distance the element moves from one view to the next.
But is this distance an objective two-dimensional measurement
or an interpreted movement in three-dimensional space? There
is some confusion in the literature about this point, since
many studies have assumed that correspondence strength is
linked to the smoothness of apparent motion (Kolers 1972), and
this is apparently more closely related to three- than to two-
dimensional distances. Ullman (1977) has however shown that
this assumption is false, and that it is the two-dimensional
distance alone that determines the correspondence.

The second part of the problem is to determine the
three-dimensional structure once the correspondence between
successive views has been established. Unless this problem is
constrained in some way, it cannot be solved, so one has to
search for reasonable assumptions on which to base the design
of one's algorithms. (This state of affairs is a common one
in the theory of visual processes, as we shall see when we
discuss the problems of stereopsis, and shape from contour).
Ullman suggested basing the interpretation on the following
assumptions; (1) any two-dimensional transformation that has a
unique interpretation as a rigid body moving in space should
be interpreted as such an object in motion, and (2) that the
imaging process is locally an orthogonal projection. He then
showed that under orthogonal projection, three-dimensional
shape and motion may be recovered from as little as three
views each showing the image of the same five points, no four
of which are coplanar. This result leads to algorithms
capable of recovering shape and motion from scenes containing
arbitrary objects in motion. The final question is whether
the algorithms that humans employ to recover shape and motion
rely on these same two assumptions, and this question is
currently under investigation. The important point here is
that for more human-like algorithms, the number of views can
be traded off against the accuracy of the computation,
decreasing the emphasis on the particular number "three".

2.2: Stereopsis

Ever since Julesz (1971) made the first random-dot
stereogram, it has been clear that at least to a first
approximation stereo vision can be regarded as a modular
component of the human visual system. Marr (1974) and Marr &
Poggio (1976) formulated the computational theory of the

stereo matching problem in the following way:

(R1) Uniqueness. Each item from each image may be assigned at most one disparity value. This condition rests on the premise that the items to be matched correspond to physical marks on a surface, and so can be in only one place at a time.

(R2) Continuity. Disparity varies smoothly almost everywhere. This condition is a consequence of the cohesiveness of matter, and it states that only a relatively small fraction of the area of an image is composed of boundaries.

By representing these constraints geometrically, Marr & Poggio (1976) embodied them in a cooperative algorithm. In figure 9, *Lx* and *Rx* represent the positions of descriptive elements from the left and right views, and the horizontal and vertical lines indicate the range of disparity values that can be assigned to left-eye and right-eye elements. The uniqueness condition then corresponds to the assertion that only one disparity value may be "on" along each horizontal or vertical line. The continuity condition states that we seek solutions that tend to spread along the dotted diagonals, which are lines of constant disparity, and between adjacent diagonals. Figure 9b shows how this geometry appears at each intersection point. Figure 9c gives the corresponding local geometry when the images are two-dimensional rather than one.

It can be shown (Marr, Poggio & Palm 1977) that, if a network is created with the positive and negative connections shown in figure 9c, states of such a network that satisfy the constraints on the computation are stable, and that given suitable inputs, the network will converge to these stable states for a wide variety of the control parameters. Thus one can think of the network as defining an algorithm that operates on many input elements to produce a global organization *via* local but highly interactive constraints.

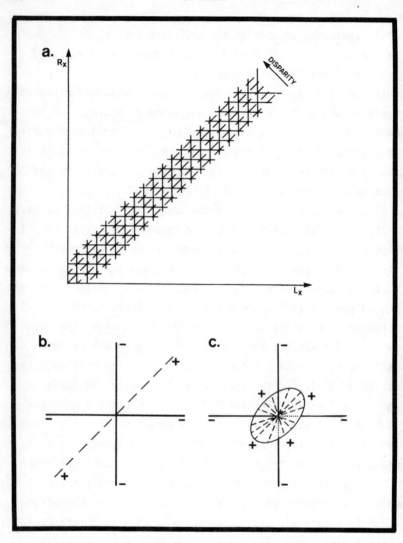

Reprinted from *Science,* Vol. 194, p. 285, October, 15, 1976.
Copyright 1976 by the American Association for the Advancement of
Science.

9.   Figure 9a shows the explicit structure of the two rules *R1*
and *R2* for the case of a one-dimensional image, and it also
represents the structure of a network for implementing the
algorithm described by equation 1.   Solid lines represent
"inhibitory" interactions, and dotted lines represent
"excitatory" ones.   9b gives the local structure at each node
of the network 9a.   This algorithm may be extended to two-
dimensional images, in which case each node in the
corresponding network has the local structure shown in 9c.
Such a network was used to solve the stereograms exhibited in
figures 10 and 11.   (Marr & Poggio 1976 figure 2).

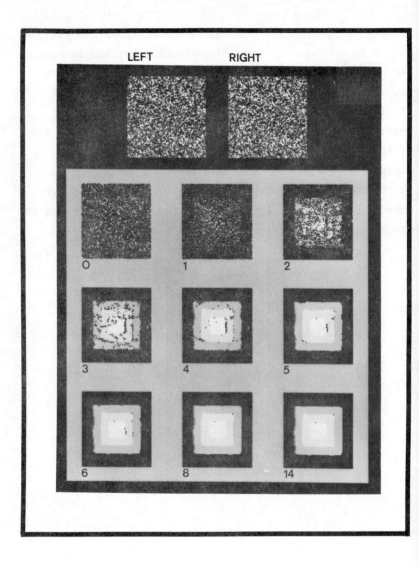

LEFT   RIGHT

0   1   2

3   4   5

6   8   14

10.   This and the following figure show the results of
applying the algorithm defined by equation (1) to two random-
dot stereograms. The initial state of the network C_{xyd} is
defined by the input such that a node takes the value 1 if it
occurs at the intersection of a 1 in the left and right eyes
(see figure 9), and it has value 0 otherwise. The network
iterates on this initial state, and the parameters used here,
as suggested by the combinatorial analysis, were θ = 3.0, ϵ =
2.0 and M = 5, where θ is the threshold and M is the
diameter of the "excitatory" neighborhood illustrated in
figure 9c. The stereograms themselves are labelled LEFT and
RIGHT, the initial state of the network as 0, and the state
after n iterations is marked as such. To understand how the
figures represent states of the network, imagine looking at it
from above. The different disparity layers in the network lie
in parallel planes spread out horizontally, so that the viewer
is looking down through them. In each plane, some nodes are
on and some are off. Each of the seven layers in the network
has been assigned a different gray level, so that a node that
is switched on in the top layer (corresponding to a disparity
of +3 pixels) contributes a dark point to the image, and one
that is switched on in the lowest layer (disparity = -3)
contributes a lighter point. Initially (iteration 0) the
network is disorganized, but in the final state, stable order
has been achieved (iteration 14), and the inverted wedding-
cake structure has been found. The density of this stereogram
is 50%. (Marr & Poggio 1976 figure 3).

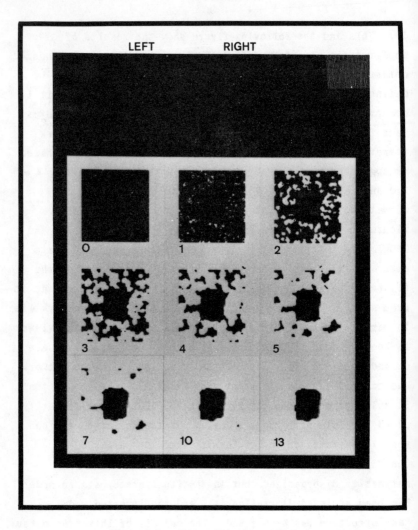

LEFT   RIGHT

Reprinted from *Science,* Vol. 194, p. 286, October, 15, 1976.
Copyright 1976 by the American Association for the Advancement of
Science.

11.   The algorithm of equation 1, with parameter values given
in the legend to figure 10, is capable of solving random-dot
stereograms with densities from 50% down to less than 10%.
For this and smaller densities, the algorithm converges
increasingly slowly.   If a simple homeostatic mechanism is
allowed to control the threshold θ as a function of the
average activity (number of "on" cells) at each iteration, the
algorithm can solve stereograms whose density is very low.   In
this example, the density is 5% and the central square has a
disparity of +2 relative to the background.   The algorithm
"fills in" those areas where no dots are present, but it takes
several more iterations to arrive near the solution than in
cases where the density is 50%.   When we look at a sparse
stereogram, we perceive the shapes in it as cleaner than those
found by the algorithm.   This seems to be due to subjective
contours that arise between dots that lie on shape boundaries.
(Marr and Poggio 1976 figure 4).

Formally, the algorithm reads:

$$C_{xyd}^{(n-1)} = u\left\{ \sum_{x'y'd' \in S(xyd)} C_{xyd}^{(n)} - \epsilon \sum_{x'y'd' \in O(xyd)} C_{x'y'd'}^{(n)} + C_{xyd}^{(o)} \right\} \qquad (1)$$

where $u(z) = 0$ if $z < \theta$, and $u(z) = 1$ otherwise; S and O are the circular and thick line neighborhoods of the cell C_{xyd} in figure 9c. This is an example of a "cooperative" algorithm (Marr & Poggio 1977a), and it exhibits typical non-linear cooperative phenomena like hysteresis, filling-in, and disorder-order transitions. Figures 10 and 11 illustrate two applications of the algorithm to random-dot stereograms.

There are a number of findings that cast doubt on the relevance of this algorithm to the question of how human stereo vision works. The most important of these findings are (a) the apparently crucial role played by eye-movements in human stereo vision (see especially Richards 1977); (b) our ability to tolerate up to 15% expansion of one image (Julesz 1971 figure 2.8.8); (c) our ability to tolerate the severe defocussing of one image (Julesz 1971 figure 3.10.3); (d) evidence that stereo detectors are organized into "three pools" (convergent, zero disparity, and divergent) and that this organization is important for stereo vision (Richards 1971); and (e) our ability to perceive depth in rivalrous stereograms (Mayhew & Frisby 1976). These difficulties led Marr & Poggio (1977b) to formulate a second stereo algorithm, designed specifically as a model for human stereopsis.

Our first stereo theory was inspired by Julesz's belief that stereoscopic fusion is a cooperative process -- a belief based primarily on the observation that it exhibits hysteresis. The main problem with the cooperative algorithm is that it apparently works too well in some ways (it performs better that humans do when eye-movements are eliminated), and

not well enough in others (humans see depth in rivalrous
stereograms). Our ability to fuse two images when one is
blurred, the rivalrous stereogram results of Mayhew & Frisby
(1976), and the recent results of Julesz & Miller (1975) on
the existence of independent spatial-frequency-tuned channels
in binocular fusion, suggest that several copies of the image,
obtained by successively coarser filtering, are used during
fusion, perhaps helping one another in a way similar to that
in which local regions help each other in our cooperative
algorithm.

The second idea was a notion that originated with Marr
& Nishihara (1977) and about which I shall have more to say
later, which is that one of the things early visual processing
does is to construct a "depth map" of the surfaces round a
viewer. In this map, each direction away from the viewer is
associated with a distance (or some function of distance) and
a surface orientation. We have christened the resulting
datastructure the $2\frac{1}{2}$-D *sketch*.

The important point here is that the $2\frac{1}{2}$-D sketch is
in some sense a memory. This provided the key idea: Suppose
that the hysteresis Julesz observed is not due to a
cooperative process at all, but is in fact the result of using
a memory buffer in which to store the depth map of the image
as it is discovered. Then, the fusion process itself need not
be cooperative, and in fact it would not even be necessary for
the whole image ever to be fused everywhere provided that a
depth map of the viewed surface were built and maintained in
this intermediate memory. This idea leads to the following
theory. (1) Each image is convolved with bar-shaped masks of
various sizes, and matching takes place between peak mask
values for disparities up to about twice the panel-width of
the mask (see Felton, Richards & Smith 1972), for pairs of

masks of the same size and polarity.   (2) Wide masks can
control vergence movements, thus causing small masks to come
into correspondence.   (3) When a correspondence is achieved
it is held and written down somewhere (e.g. in the $2\frac{1}{2}$-D
sketch).   (4) There is a backwards relation between the memory
and the masks, perhaps simply through the control of eye-
movements, that allows one to fuse any piece of a surface
easily once its depth map has been established in the memory.

This theory leads to many experimental predictions,
which are currently being tested.

3: Intermediate processing problems

3.0 Introduction

We have discussed the types of information that need to be represented early in the processing of visual information, and we have examined the computational structure of some of the processes that can derive and maintain this information. We turn now to the question of what all this information is to be used for.

3.1 Difficulties with the idea of image segmentation

The current approach to machine vision assumes that the next step in visual processing consists of a process called *segmentation*, whose purpose is to divide the image into regions that are meaningful either in terms of physical objects or for the purpose at hand. Despite considerable efforts over a long period, the theory and practise of segmentation remain primitive, and once again I believe that the main reason lies in the failure to formulate precisely the goals of this stage of the processing. What for example is an object? Is a head one? Is it still one if it is attached to a body? What about a man on horseback?

These questions point to some of the difficulties one has when trying to formulate what should be recovered as a region from early visual processing. Furthermore, however one chooses to answer them, it is usually still impossible to recover the desired regions using only local grouping

techniques acting on a representation like the primal sketch.
Most images are too complex, and even the simplest images
cannot often be segmented entirely at that level (e.g. Marr
1976 figure 13).

Something additional is clearly needed, and one
approach to the dilemma has been to invoke specialized
knowledge about the nature of the scenes being viewed to aid
segmentation of the image into regions that correspond roughly
to the objects expected in the scene. Tenenbaum & Barrow
(1976), for example, applied knowledge about several different
types of scene to the segmentation of images of landscapes, an
office, a room, and a compressor. Freuder (1974) used a
similar approach to identify a hammer in a simple scene. If
this approach were correct, it would mean that a central
problem for vision is arranging for the right piece of
specialized knowledge to be made available at the appropriate
time during segmentation. Freuder's work, for example, was
almost entirely devoted to the design of a heterarchical
control system that made this possible. More recently, the
constraint relaxation technique of Rosenfeld, Hummel & Zucker
(1976) has attracted considerable attention for just this
reason, that it appears to offer a technique whereby
constraints drawn from disparate sources may be applied to the
segmentation problem whilst incurring only minimal penalties
in control. It is however difficult to analyze such
algorithms rigorously even in very clearly defined situations
(see e.g. Marr, Poggio & Palm 1977), and in the naturally
more diffuse circumstances that surround the segmentation
problem, it may often be impossible.

3.2: *Reformulating the problem*
The basic problem seems to be how to formulate

precisely the next stage of visual processing. Given a
representation like the primal sketch, and the many possible
boundary-defining processes that are naturally associated with
it, which boundaries should one attend to and why? The
segmentation approach fails because objects and desirable
regions are not visually primitive constructions, and hence
cannot be recovered reliably from the primal sketch or similar
representation without additional specialized knowledge. If
we are to succeed, we must discover precisely what information
it is that needs to be made explicit at this stage, what, if
any, additional knowledge it is appropriate to apply, and we
must design a representation that matches these requirements.

In order to search for clues to a suitable
representation, let us return to the physics of the situation.
The primal sketch represents intensity changes and the local
two-dimensional geometry of an image. The principle factors
that determine these are (1) the illuminant, (2) surface
reflectance, (3) the shape of the visible surface, and (4) the
vantage point. The first two factors raise the difficult
problems of color and brightness, and I shall not discuss them
further. The third and fourth factors are independent of the
first two (whether two shapes are the same does not depend
upon their colors or on the lighting), and so may be treated
separately.

I shall argue that, since most early visual processes
extract information about the visible surface, it is these
surfaces, their shape and disposition relative to the viewer,
that need to be made explicit at this point in the processing.
Furthermore, because surfaces exist in three-dimensional
space, this imposes constraints on them that are general, and
not confined to particular objects. It is these constraints
that constitute the *a priori* knowledge that it is appropriate

to bring to bear next.

One example of the exploitation of fairly general
constraints was the work of Waltz (1975), who formulated the
constraints that apply to images of polyhedra. The
representation on which that work was based was line drawings,
but these are not suitable for our needs here, because part of
the task we wish to carry out is the discovery of physical
edges that are only weakly present or even absent in the
primal sketch. The approach of Mackworth (1973) was closer to
what we want, since it involved a primitive way of
representing surfaces.

3.3: General classification of shape representations

Part of our task in formulating the problem of
intermediate vision is therefore the examination of ways of
representing and reasoning about surfaces. We therefore start
our enquiry by discussing the general nature of shape
representations. What kinds are there, and how may one decide
among them? Although it is difficult to formulate a
completely general classification of shape representations,
Marr & Nishihara (1977) attempted to set out the basic design
choices that have to be made when a representation is
formulated. They concluded that there are three
characteristics of a shape representation that are largely
responsible for determining the information that it makes
explicit.   The first is the type of *coordinate system* it uses,
whether it is defined relative to the viewer or to the object
being viewed; the second characteristic concerns the nature of
the *shape primitives* used by the representation, that is, the
elements whose positions the coordinate system is used to
define.   Are they two- or three-dimensional, in what sizes do
they come, and how detailed are they? And the third is

concerned with the organization a representation imposes on
the information in a description, for example is the
description modular or does it have little internal structure?
We have two sources of information that can help us to
formulate the important issues in intermediate visual
information processing, firstly the computational problems
that arise, and secondly, psychophysics.

3.4: Some observations from psychophysics

Vision provides several sources of information about
shape. The most direct are stereo and motion, but texture
gradients in a single image are nearly as effective, and the
theatrical techniques of facial make-up rely on the
sensitivity of perceived shape to shading. It often happens
that some parts of a scene are open to inspection by some of
these techniques, and other parts by others. Yet different as
the techniques are, they have two important characteristics in
common. They rely on information from the image rather than
on *a priori* knowledge about the shapes of the viewed objects;
and the information they specify concerns the depth or surface
orientation at arbitrary points in an image, rather than the
depth or orientation associated with particular objects.

If one views a stereo pair of a complex surface, like
a crumpled newspaper or the "leaves" cube of Ittelson (1960),
one can easily state the surface orientation of any piece of
the surface, and whether one piece is nearer to or further
from the viewer than its neighbors. Nevertheless one's memory
for the shape of the surface is poor, despite the vividness of
its surface orientation during perception. Furthermore, if
the surface contains elements nearly parallel to the line of
sight, their apparent surface orientation when viewed
monocularly can differ from the apparent surface orientation

when viewed binocularly.

From these observations, one can perhaps draw some
simple inferences.
(a) There is at least one internal representation of the
depth, or surface orientation, or both, associated with each
surface point in a scene.
(b) Because surface orientation can be associated with
unfamiliar shapes, its representation probably precedes the
decomposition of the scene into objects. (This point is
particularly relevant to our discussion of intermediate visual
information processing.)
(c) Because the apparent orientation of a surface element can
change, depending on whether it is viewed binocularly or
monocularly, the representation of surface orientation is
probably driven almost entirely by perceptual processes, and
is influenced only slightly by specific knowledge of what the
surface orientation actually is. Our ability to "perceive"
the surface much better than we can "memorize" it may also be
connected with this point.
In addition, it seems likely that the different sources of
information can influence the *same* representation of surface
orientation.

3.5: The computational problem
In order to make the most efficient use of these
different and often complementary sources of information, they
need to be combined in some way. The computational question
is, how best to do this? The natural answer is to seek some
representation of the visual scene that makes explicit just
the information these processes can deliver.
Fortunately, the physical interpretation of the

Table 1

The form in which various early visual processes deliver information about the changes in a scene.

r = depth
δr = small, local changes in depth
Δr = large changes in depth
$\underline{s}$ = local surface orientation

Information source	Natural parameter
Stereo	Disparity, hence especially δr and Δr
Motion	r, hence δr, Δr
Shading	$\underline{s}$
Texture gradients	$\underline{s}$
Perspective cues	$\underline{s}$
Occlusion	Δr

representation we seek is clear. All these processes deliver
information about the depth or surface orientation associated
with surfaces in an image, and these are well-defined physical
quantities. We therefore seek a way of making this
information explicit, of maintaining it in a consistent state,
and perhaps also of incorporating into the representation any
physical constraints that hold for the values that depth and
surface orientation take over the kinds of surface that occur
in the real world. Table 1 lists the type of information that
the different early processes can extract from images. The
interesting point here is that although processes like stereo
and motion are in principle capable of delivering depth
information directly, they are in practise more likely to
deliver information about local *changes* in depth, for example
by measuring local changes in disparity. Texture gradients
and shading provide more direct information about surface
orientation. In addition, occlusion and brightness and size
clues can deliver information about discontinuities in depth.
(It is for example amazing how clear an impression of depth
can be obtained from a monocular image containing bright or
dim rectangles of different sizes against a dark background).
The main function of the representation we seek is therefore
not only to make explicit information about depth, local
surface orientation, and discontinuities in these quantities,
but also to create and maintain a global representation of
depth that is consistent with the local cues that these
sources provide. We call such a representation the $2\frac{1}{2}$-D
sketch, and the next section describes a particular candidate
for it.

3.6: A possible form for the $2\frac{1}{2}$-D sketch
        The example I give for the $2\frac{1}{2}$-D sketch is a viewer-

centered representation, which uses surface primitives of one
(small) size. It includes a representation of contours of
surface discontinuity, and it has enough internal
computational structure to maintain its descriptions of depth,
surface orientation and surface discontinuity in a consistent
state. The representation itself has no additional internal
structure.

Depth may be represented by a scalar quantity r, the
distance from the viewer of a point on a surface. Surface
discontinuities may be represented by oriented line elements.
Surface orientation may be represented by a unit vector (x, y, z) in three-dimensional space. Following those who have used
gradient space (Huffman 1971, Horn 1977) we can rewrite this
as $(p, q, 1)$, which can be represented as a vector (p, q) in
two-dimensional space. In other words, surface orientation
may be represented by covering an image with needles. The
length of each needle defines the dip of the surface at that
point, so that zero length corresponds to a surface that is
perpendicular to the vector from the viewer to that point, and
the length increases as the surface tilts away from the
viewer. The orientation of the needle defines the direction
of the surface's dip. Figure 12 illustrates this
representation.

In principle, the relation between depth and surface
orientation is straightforward -- one is simply the integral
of the other, taken over regions bounded by surface
discontinuities. It is therefore possible to devise a
representation with intrinsic computational facilities that
can maintain the two variables, of depth and surface
orientation, in a consistent state. But note that, in any
such scheme, *surface discontinuities* acquire a special status
(as curves across which integration stops). Furthermore, if

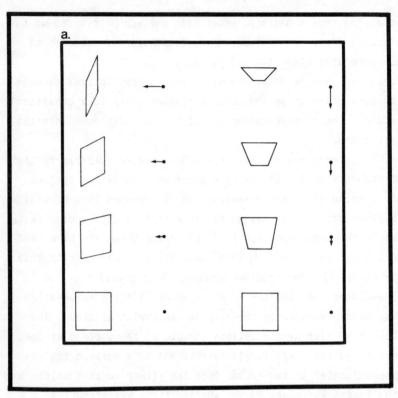

Reprinted with permission of the Royal Society.

12.   The $2\frac{1}{2}$-D sketch represents depth, contours of surface
discontinuity, and the orientation of visible surfaces.   A
convenient representation of surface orientation is described
in the text and illustrated here. The orientation of the
needles is determined by the projection of the surface normal
on the image plane, and the length of the needles represents
the dip out of that plane (12a).   A typical $2\frac{1}{2}$-D sketch
appears in 12b, although depth information is not represented
in the figure.   (Marr & Nishihara 1977 figure 2).

b.

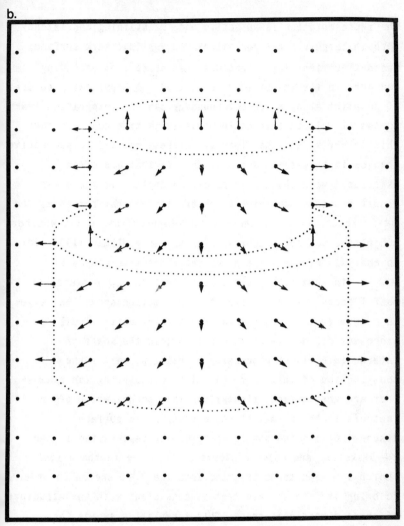

Reprinted with the permission of Springer-Verlag, publishers.

the representation is an active one, maintaining consistency
through largely local operations, curves that mark surface
discontinuities (e.g. contours that arise from occluding
contours in the image) must be "filled in" completely, so that
at no point along an object boundary can the integration leak
across it.   It is interesting that subjective contours have
this property, and that they are closely related to subjective
changes in brightness (cf section 1.6) that are often
associated with changes in perceived depth.   If the human
visual processor contains a representation that resembles the
$2\frac{1}{2}$-D sketch, it would therefore be interesting to ask whether
subjective contours occur within it.   (See Ullman (1976) for
an analysis of the shape of curved subjective contours).

In summary, my argument is that the $2\frac{1}{2}$-D sketch is
useful because it makes explicit information about the image
in a form that is closely matched to what early visual
processes can deliver.   We can formulate the goals of
intermediate visual processing as being primarily the
construction of this representation, discovering for example
what are the surface orientations in a scene, which of the
contours in the primal sketch correspond to surface
discontinuities and should therefore be represented in the
$2\frac{1}{2}$-D sketch, and which contours are missing in the primal
sketch and need to be inserted into the $2\frac{1}{2}$-D sketch in order
to bring it into a state that is consistent with the structure
of three-dimensional space. This formulation avoids the
difficulties associated with the terms "region" and "object",
and allows one to ask precise questions about the
computational structure of the $2\frac{1}{2}$-D sketch and of processes
to create and maintain it.   We are currently much occupied
with these problems.

4: Later processing problems

4.0: Introduction

The $2\frac{1}{2}$-D sketch is a poor representation for the purposes of recognition because it is unstable (in the sense of Marr & Nishihara 1977), it depends on the vantage point, and it fails to make explicit pieces of a shape (like an arm) that are larger that the primitive size. Except for the simplest of purposes, it is an inadequate vehicle for a visual system to convey information about shape to other processes, and so I turn now to representations that are more suitable for recognition tasks.

If one were to design a shape representation to suit the problems of recognition, one would naturally base it on an object-centered coordinate system. In addition, one would have to include shape primitives of many different sizes, so as to be able to make explicit shape characteristics that can range from a wart to an elephant. Marr & Nishihara (1977) discuss these questions in detail, and I shall not repeat their observations here. The deepest issues are those raised by having to define an object-based coordinate system. Since they are central to the problem of defining representations for use in later processing of visual information, I shall spend the remainder of the essay discussing this topic.

4.1: Nature of an object-centered coordinate system

Marr & Nishihara (1977) pointed out that there are two

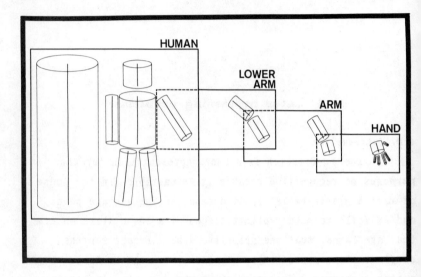

Reprinted with the permission of Springer-Verlag, publishers.

13.   This diagram illustrates the organization of shape
information in a 3-D model description.   Each box corresponds
to a 3-D model; with its model axis on the left side of the
box and the arrangement of its component axes are shown on the
right side.   In addition some component axes have 3-D models
associated with them and this is indicated by the way the
boxes overlap.   The relative arrangement of each model's
component axes, however, is shown improperly since it should
be in an object-centred system rather than the viewer-centred
projection used here.   This example shows a coarse overall
description of a human shape along with an elaboration of one
of its components (the arm).   The important characteristics of
this type of organization are: (i) each 3-D model is a self-
contained unit of shape information and has a limited
complexity, (ii) information appears in shape contexts
appropriate for recognition (the disposition of a finger is
most stable when specified relative to the hand that contains
it), and (iii) the representation can be used flexibly
(components can be elaborated according to the needs of the
moment or the time available, and a 3-D model description of a
component is easily added to a description of the whole
shape).   The major limitation imposed on the representation by
this form of oraganization is on its scope, since it will only
be useful for shapes for which the decomposition into 3-D
models is well defined.   (Marr & Nishihara 1977 figure 3).

types of object-centered coordinate system that one might
attempt to define precisely. One refers all locations on an
object to a single coordinate frame that embraces the entire
object, and the other distributes the coordinate system,
making it local to each articulated component or individual
shape characteristic. Marr & Nishihara concluded that the
second of these schemes is the more desirable, and they gave
as an example the representation illustrated in figure 13.
But with a representation of this kind, the most difficult
questions begin after its internal structure has been defined.
How can one define canonically the coordinate scheme for an
arbitrary shape, and even more difficult, how can such a thing
be found from an image *before* a description of the viewed
shape has been computed? Some kind of answers to these
questions must be found if the representation is to be used
for recognition.

4.2: *Shapes having natural coordinate systems*

If the coordinate system used for a given shape is to
be canonical, its definition must take advantage of any
salient geometrical characteristics that the shape possesses.
For example, if a shape has natural axes, distinguished by
length or by symmetry, then they should be used. The
coordinate system for a sausage should take advantage of its
major axis, and for a face, of its axis of symmetry.

Highly symmetrical objects, like a sphere, square, or
circular disc, will inevitably lead to ambiguities in the
choice of coordinate systems. For a shape as regular as a
sphere this poses no great problem, because its description in
all reasonable systems is the same. One can even allow other
factors, like the direction of motion or of spin, to influence
the choice of coordinate frame. For other shapes, the

existence of more than one possible choice probably means that
one has to represent the object in several ways. This is
acceptable provided that the number of ways is small. For
example, there are four possible axes on which one might wish
to base the coordinate system for representing a door, the
midlines along its length, its width, its thickness, and to
represent how the door opens, the axis of its hinges. For a
typewriter, there are two choices at the top level; an axis
parallel to its width, because that is usually its largest
dimension, and the axis about which a typewriter is roughly
symmetrical.

In general, if an axis can be distinguished in a
shape, it can be used as the basis for a local coordinate
system. One approach to the problem of defining object-
centered coordinate systems is therefore to examine the class
of shapes having an axis as an integral part of their
structure. One such is the class of *generalized cones*. (A
generalized cone is the surface swept out by moving a cross
section of constant shape but smoothly varying size along an
axis, as in figure 14). Binford (1971) drew attention to this
class of surfaces, suggesting that it might provide a
convenient way of describing three-dimensional surfaces for
the purposes of computer vision. I regard it as an important
class not because the shapes themselves are easily decribable,
but because the presence of an axis allows one to define a
canonical local coordinate system. Fortunately many objects,
especially those whose shape was achieved by growth, are
described quite naturally in terms of one or more generalized
cones. The animal shapes in figure 15 provide some examples
-- the individual sticks are simply axes of generalized cones
that approximate the shapes of parts of these animals. Many
artifacts can also be described in this way, like a car (a

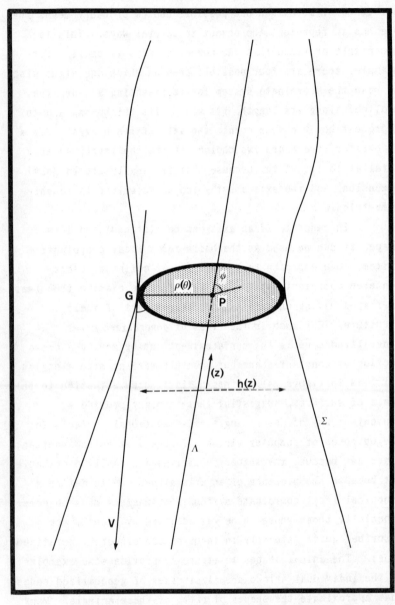

Reprinted from D. Marr, *Analysis of occluding contour*, Proceedings of the Royal Society of London, Ser. B., vol. 197 (1977) p. 447.

14.   The definition of a generalized cone. In this article, a generalized cone is the surface generated by moving a smooth cross-section ρ along a straight axis Λ. The cross-section may vary smoothly in size (as prescribed by the function $h(z)$), but its shape remains constant. The eccentricity of the cone is the angle ψ between its axis and a plane containing a cross-section.   (Figure 5 of Marr 1977).

Reprinted with the permission of Springer-Verlag, publishers.

15. These pipecleaner figures illustrate the point that a
shape representation does not have to reproduce a shape's
surface in order to describe it adequately for recognition; as
we see here, animal shapes can be portrayed quite effectively
by the arrangement and relative sizes of a small number of
sticks. The simplicity of these descriptions is due to the
correspondence between the sticks shown here and natural or
canonical axes of the shapes described. To be useful for
recognition, a shape representation must be based on
characteristics that are uniquely defined by the shape and
which can be derived reliably from images of it. (Marr &
Nishihara 1977 figure 1).

158   D. MARR

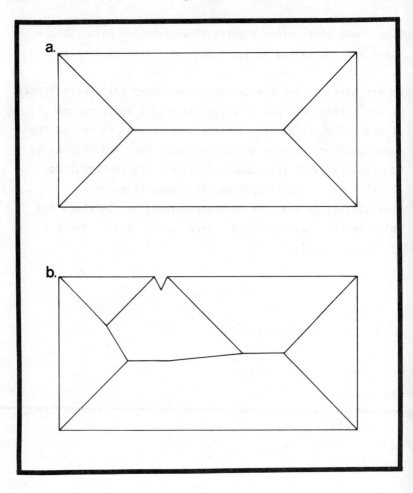

16.   Blum's (1973) grassfire technique for recovering an axis
from a silhouette is undesirably sensitive to small
perturbations in the contour. 16a shows the Blum transform of
a rectangle, and 16b, of a rectangle with a notch. (Redrawn
from Agin 1972).

small box sitting atop and in the middle of a longer one), and a building (a box with a vertical axis).

It is important to remember that there exist surfaces that cannot conveniently be approximated by generalized cones, for example a cake that has been cut at its intersection with some arbitrary plane, or the surface formed by a crumpled newspaper. Cases like the cake can be dealt with by introducing suitable surface primitives that describe the plane of the cut, but the crumpled newspaper poses apparently intractable problems.

4.3: Finding the natural coordinate system from an image

Even if a shape possesses a canonical coordinate system, one is still faced with the problem of finding it from an image. Blum (1973), Agin (1972) and Nevatia (1974) have addressed problems that are related to this question. Blum's sym-axis theory is an interesting one, because he specifies precisely what it is that is computed from a two-dimensional outline. Unfortunately, it is not clear that what this theory computes is in fact useful for shape recognition (see e.g. figure 16), and when applied to a three-dimensional shape, the sym-axis is in general a two-dimensional sheet, so it cannot easily be used to define an object-centered coordinate system. Agin's and Nevatia's work, on the other hand, concerns the analysis of a depth map. This is an important problem, and it would be interesting to see a careful analysis of the conditions under which their techniques will succeed.

My own interest in the problem grew from the 3-D representation theory of Marr & Nishihara (1977), in particular from the question of how to interpret the outlines of objects as seen in a two-dimensional image. The rest of this essay summarizes a recent article by Marr (1977). The

17.   "Rites of spring" by P. Picasso. We immediately interpret
the silhouettes in terms of particular 3-D surfaces, despite
the paucity of information in the image.   In order to do this,
we must be bringing additional assumptions and constraints to
bear on the analysis of these contours' shapes.   Marr (1977)
enquired about the nature of this *a priori* information.

starting point for this work was the observation that when one
looks at the silhouettes in Picasso's work "Rites of Spring"
(figure 17), one perceives them in terms of very particular
three-dimensional shapes, some familiar, some less so. This
is quite remarkable, because the silhouettes could in theory
have been generated by an infinite variety of shapes which,
from other viewpoints, have no discernable similarities to the
shapes we perceive.   One can perhaps attribute part of the
phenomenon to a familiarity with the depicted shapes; but not
all of it, because one can use the medium of a silhouette to
convey a new shape, and because even with considerable effort
it is difficult to imagine the more bizarre three-dimensional
surfaces that could have given rise to the same silhouettes.
The paradox is, that the bounding contours in figure 17
apparently tell us more than they should about the shape of
the dark figures.   For example, neighboring points on such a
contour could in general arise from widely separated points on
the original surface, but our perceptual interpretation
usually ignores this possibility.

     The first observation to be made here is that the
occluding contours that bound these silhouettes are contours
of surface discontinuity, that is precisely the contours with
which the $2\frac{1}{2}$-D sketch is concerned.   Secondly, because we can
interpret them as three-dimensional shapes, then implicit in
the way we interpret them must lie some *a priori* assumptions
that allow us to infer a shape from an outline.   If a surface
violates these assumptions, our analysis will be wrong, in the
sense that the shape we assign to the contours will differ
from the shape that actually caused them. An everyday example
of this phenomenon is the shadowgraph, where the appropriate
arrangement of one's hands can, to the surprise and delight of
a child, produce the shadow of an apparently quite different

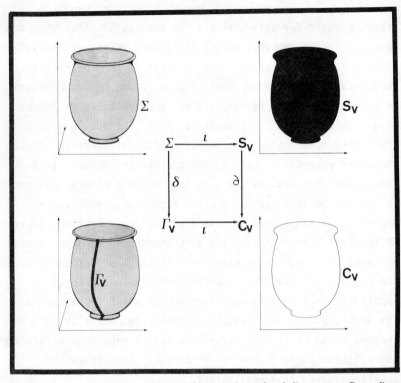

Reprinted from D. Marr, *Analysis of occluding contour,* Proceedings
of the Royal Society of London, Ser. B., vol. 197 (1977) p. 444.

18.   From viewpoint V, the three-dimensional surface Σ forms
the silhouette S_V in the image *via* the imaging process ι.   The
boundary of S_V, obtained by the boundary operator ∂ is denoted
by C_V and we call it the contour of Σ.   The set of points on
Σ that ι maps onto C_V we call the contour generator of C_V, and
it is denoted by Γ_V.   The map from Σ to Γ_V induced by ∂ is
denoted by δ.   (Figure 2 of Marr 1977).

shape, like a duck or a rabbit.

What assumptions is it reasonable to suppose that we make? In order to explain them, I need to define the four structures that appear in figure 18. These are (1) some three dimensional surface Σ; (2) its image or silhouette S_V as seen from a viewpoint V; (3) the bounding contour C_V of S_V; and (4) the set of points on the surface Σ, that project onto the contour C_V. We shall call this last set the *contour generator* of C_V, and we shall denote it by Γ_V.

If one is presented with a contour in an image, without any knowledge of the surface or perspective that caused it, there is very little information on which one can base one's analysis. The only obvious feature available is the distinction between convex and concave pieces of contour -- that is, the presence of inflection points. In order that inflection points be "reliable", one needs to make some assumptions about the way the contour was generated, and I chose the following restrictions:

R1: The surface Σ is smooth.

R2: Each point on the contour generator Γ_V projects to a different point on the contour C_V.

R3: Nearby points on the contour C_V arise from nearby points on the contour generator Γ_V.

R4: The contour generator Γ_V of C_V is planar.

The first restriction is only a technical one. The second and third say that each point on the contour in the image comes from one point on the surface (which is an assumption that facilitates the analysis but is not of fundamental importance), and that where the surface looks continuous in the image, it really is continuous in three

dimensions. The fourth condition, together with the
constraint that the imaging process be an orthogonal
projection, is simply a necessary and sufficient condition
that the difference between convex and concave contour
segments reflects properties of the surface, rather than
characteristics of the imaging process.

It turns out that the following theorem is true, and
it is a result that I found very surprising.

Theorem. If *R1* is true, and *R2 - R4* hold for all
distant viewing directions that lie in some plane,
then the viewed surface is a generalized cone.

This means that if, for distant viewpoints whose
viewing directions lie parallel to some plane, a surface's
shape can successfully be inferred using only the convexities
and concavities of its bounding contours in an image, then
that surface is a generalized cone or is composed of several
such cones. The interesting thing about this result is that
it implicates generalized cones. We have already seen that
the important thing about these cones is that an axis forms an
integral part of their structure. But this is a feature of
their three-dimensional organization, and ought in some sense
to be independent of the issues raised by vision. What the
theorem says is that there is a natural link between
generalized cones and the imaging process itself. The
combination of these two must mean, I think, that generalized
cones will play an intimate role in the development of vision
theory.

4.4: Interpreting the image of a single generalized cone
If we take this result at face value, we can now ask

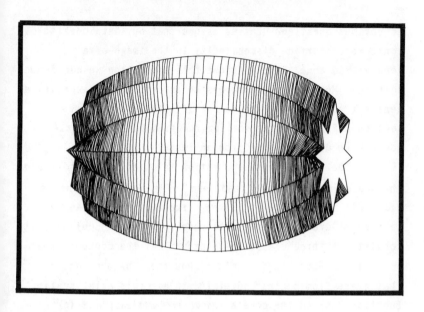

19.   A sketch of a generalized cone showing its silhouette
(the circumscribing contour), and its fluting (the contours
spanning its length). The radial extremeties of a generalized
cone are illustrated in figure 20.

an obvious question. Let us assume that our data consist of
contours of surface discontinuity in the image of a
generalized cone, since without this assumption we can deduce
nothing. How may such contours be interpreted? To specify a
generalized cone, we have to specify its axis Λ, cross-
section $\rho(\theta)$, and axial scaling function $h(z)$ (figure 14);
how can we discover them from an image?

The answer to this question is based on the notion of
the *skeleton* of a generalized cone. The skeleton is not a
difficult idea, since it is very like the set of lines a
cartoonist draws to convey the shape of a curved object. It
consists of three classes of contour: (a) the contours that
occur in a generalized cone's silhouette; (b) the contours
that arise from maxima and minima in a cone's axial scaling
function (called the cone's *radial extremities*); and (c)
contours that arise from maxima and minima in the cone's
cross-section (its *fluting*). These categories are illustrated
in figure 19.

The reason why the skeleton is a useful construct for
recognition is that one can detect its presence in an image by
the many relationships that exist among its parts. For
example, radial extremities are all parallel to each other,
and the silhouette and fluting have a kind of symmetry about
the image of the cone's axis. It turns out that one can use
these relationships to set up constraints on a set of contours
such that, if those constraints are all satisfied by a unique
interpretation of the contours in the image, one can be
reasonably certain that a skeleton has been found, and hence
that the contours can be interpreted as arising from a
generalized cone whose axis is then determined. The practical
importance of this result is illustrated in figure 20, where
one can see that the image of the "sides" is symmetrical about

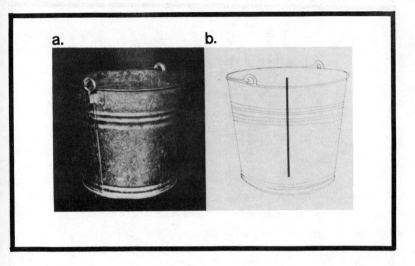

Reprinted from D. Marr, *Analysis of occluding contour,* Proceedings of the Royal Society of London, Ser. B., vol. 197 (1977) p. 459.

20.   Methods based on the theory described in the text suffice to solve this image of a bucket.   An axial symmetry is established by its sides about the bucket's axis (shown thickened), and a parallel relationship holds between components of its radial extremity.   (Figure 14 of Marr 1977).

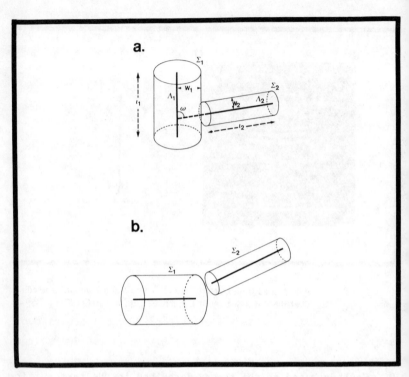

Reprinted from D. Marr, *Analysis of occluding contour,* Proceedings
of the Royal Society of London, Ser. B., vol. 197 (1977) p. 459.

21.   The two main types of joins between two generalized
cones.   21a shows a side-to-end join, and 21b shows an end-to-
end join. (Figure 14 of Marr 1977).

the bucket's axis, and there is a clear parallel relationship
between the image of the bucket's top, the corrugations in its
side, and the visible part of its base (the bucket's radial
extremities).   These relations, of symmetries and
parallelism, are preserved by an orthogonal projection.   Hence
provided that the contours are formed along a viewing
direction that is not too close to the axis of the cone, these
relations will still be present in the image.   If the viewing
direction lies so close to the cone's axis that its image is
substantially foreshortened, these relationships will no
longer be present, but it is part of the overall theory that
such views have to be handled differently (Marr & Nishihara
1977).

4.5: Surfaces composed of two or more generalized cones

Real-life objects are often approximately composed of
several different cones, joined together in various ways (see
figure 13), and we therefore have to study ways of decomposing
a multiple cone into its components -- for example, a human
body into arms, legs, torso and head.   Marr (1977) analyzed
the two types of join shown in figure 21, giving criteria that
define segmentation points on the contour produced by two
joined cones (theorems 7 and 8).   Figure 22 exhibits the
segmentation points *P* and *Q* for the case in which two short
cones are joined side-to-end.   P. Vatan has written a computer
program that can carry out this segmentation, and an example
of its operation is illustrated in figure 23.   The legend to
the figure describes the particular algorithm used.

4.6: Some comments on the limitations of this theory

The results of this theory are limited in their scope
to a particular class of views and surfaces, but on the other

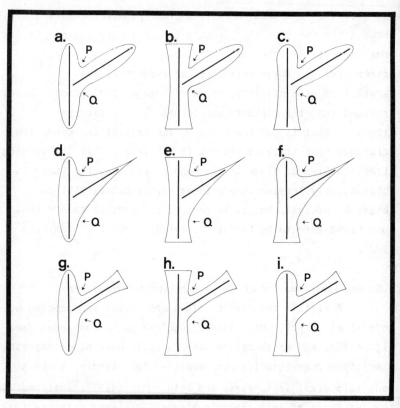

Reprinted from D. Marr, *Analysis of occluding contour,* Proceedings of the Royal Society of London, Ser. B., vol. 197 (1977) p. 461.

22. This figure illustrates the types of side-to-end join that can occur between two short generalized cones. In the first column, the left-hand cone is convex; in the center column it is concave, and in the third column, it is convex on one side of the join, and concave on the other. The other cone is convex in the top row, and concave in the other two. Segmentation depends upon finding the points *P* and *Q*, which are defined by theorem 7 of Marr (1977) and illustrated here for each case. (Figure 18 of Marr 1977).

hand, they use only a limited kind of visual information,
little more than occluding contours that are formed in an
image by rays that graze a smooth surface. Interestingly,
these particular contours are unsuitable for use in stereopsis
or structure-from-motion computations, because they are not
formed from markings that define precise locations on the
viewed surface.   Creases and folds on a surface also give rise
to contours in an image, and these have yet to be studied in
detail.   Information about shape from shading, texture, stereo
or motion information has not yet been considered.   By adding
these other sources of information, I hope that a set of
methods can eventually be assembled that together approach a
comprehensive treatment of possible image configurations.

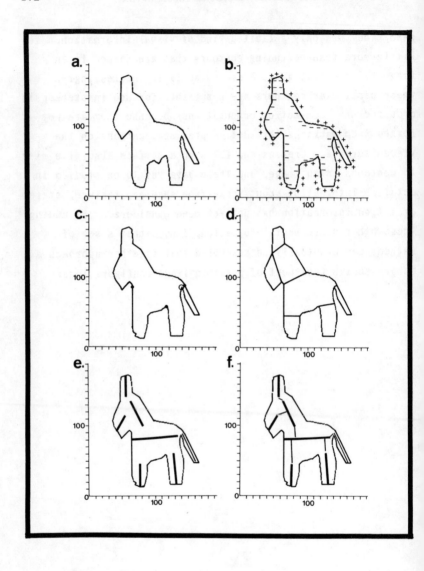

23.   The occluding contours in an image can be used to locate
the images of the natural axes of a shape composed of
generalized cones (Marr 1977).   The initial outline in (a) was
obtained by applying local grouping processes to the *primal
sketch* of the image of a toy donkey (Marr 1976).   This outline
was then smoothed and divided into convex and concave sections
to get (b).   Next, strong segmentation points, like the deep
concavity circled in (c), are identified and a set of
heuristic rules are used to connect them with other points on
the contour to get the segmentation shown in (d).   The
component axes shown in (e) are then derived from these.   The
resulting segments are checked to see that they obey the rules
for images of generalized cones.   The boundaries must for
example be symmetric about the axes, and in the case of side-
to-end joins, the axis of the cone that is attached by its end
must intersect the segmentation points that separate the two
cones' contours.   In this example, most of the symmetry
relations have degenerated into parallelism.   The thin lines
in (f) indicate the position of the head, leg, and tail
components along the torso axis, and the snout and ear
components along the head axis.   (This algorithm is due to P.
Vatan).

5: Discussion

I have tried to make three main points in this
article: The first is methodological, namely that it is
important to be very clear about the nature of the
understanding we seek (Marr & Poggio 1977a, Marr 1977b). The
results we try to achieve should be precise, at the level of
what I called a computational theory, and should deal with
problems that can confidently be attributed to a real aspect
of vision, and not (for example) to an artifact of the
limitations of one's current vision program.

The second main point is that the critical issues for
vision seem to me to revolve around the nature of the
representations used - that is, the particular characteristics
of the world that are made explicit - and the nature of the
processes that recover these characteristics, create and
maintain the representations, and eventually read them. By
analyzing the spatial aspects of the problem of vision (Marr &
Nishihara 1977), an overall framework for visual information
processing is suggested, that consists of three principal
representations: (1) the primal sketch, which makes explicit
the intensity changes and local two-dimensional geometry of an
image; (2) the $2\frac{1}{2}$-D sketch, which is a viewer-centered
representation of the depth and orientation of the visible
surfaces and includes contours of discontinuities in these
quantities; and (3) the 3-D model representation, whose
important features are (a) that its coordinate system is
object-centered, (b) that it includes volumetric primitives,

that make explicit the space occupied by an object and not just its visible surfaces, and (c) that primitives of various sizes are included, arranged in a modular, hierarchical organization.

The third main point concerns the study of processes for recovering the various aspects of the physical characteristics of a scene from images of it. The critical act in formulating computational theories for such processes is the discovery of valid constraints on the way the world behaves, that provide sufficient additional information to allow recovery of the desired characteristic. Several examples are already available, including Land & McCann (1971), which rests on the distinction between sharp and shallow intensity changes; stereopsis (Marr 1974, Marr & Poggio 1976, 1977b) which uses continuity and uniqueness; structure from visual motion (Ullman 1977), which uses rigidity; fluorescence (Ullman 1976a); and shape from contour (Marr 1977a). The discovery of constraints that are valid and sufficiently universal leads to results about vision that have the same quality of permanence as results in other branches of science (Marr 1977b).

Finally, once a computational theory for a process has been formulated, algorithms for implementing it may be designed, and their performance compared with that of the human visual processor. This allows two kinds of result. Firstly, if performance is essentially identical, one has good evidence that the constraints of the underlying computational theory are valid and may be implicit in the human processor; and secondly, if a process matches human performance, it is probably sufficiently powerful to form part of a general purpose vision machine.

Acknowledgements: I thank the members of the M.I.T.
Artificial Intelligence Laboratory vision group for making
this survey possible, especially Keith Nishihara with whom
much of the overall framework was developed, Dr. Tomaso
Poggio, Prof. Whitman Richards and Dr. Shimon Ullman for many
stimulating discussions, and Karen Prendergast for preparing
the illustrations. *Science* kindly gave permission for the
reproduction of figures 9, 10 and 11, and the Royal Society
for figures 1, 12, 13, 14, 15, 17, 18, 20, 21 and 22. This
work was conducted at the Artificial Intelligence Laboratory,
a Massachusetts Institute of Technology research program
supported in part by the Advanced Research Projects Agency of
the Department of Defense, and monitored by the Office of
Naval Research under contract number N00014-75-C-0643.

6: References

Agin, G. J. (1972) Representation and description of curved objects. *Stanford A.I. Memo 173.*

Beck, J. (1974). *Surface color perception.* Ithaca, N. Y.: Cornell University Press.

Binford, T. O. (1971) Visual perception by computer. Presented to the IEEE Conference on Systems and Control, Miami, December.

Blum, H. (1973) Biological shape and visual science, (part 1). *J. theor. Biol.* , *38,* 205-287.

Campbell, F. W. C. (1977) Sometimes a biologist has to make a noise like a mathematician. *Neurosciences Res. Prog. Bull. 15,* 417-424.

Colas-Baudelaire, P. (1973) Digital picture processing and psychophysics: a study of brightness perception. *Report No. UTEC-CSC-74-025 from the Department of Computer Science, University of Utah.*

Forbus, K. (1977) Light source effects. *M.I.T. A.I. Lab. Memo 422.*

Freuder, E. C. (1974) A computer vision system for visual recognition using active knowledge. *M.I.T. A.I. Lab. Technical Report 345.*

Glass, L. (1969) Moire effect from random dots. *Nature, 243,* 578-580.

Gilchrist, A. L. (1977) Perceived lightness depends on perceived spatial arrangement. *Science, 195,* 185-187.

Horn, B. K. P. (1974) Determining lightness from an image. *Computer Graphics and Image Processing, 3,* 277-299.

Horn, B. K. P. (1977) Image intensity understanding. *Artificial Intelligence 8,* 201-231.

Hubel, D. H. & Wiesel, T. N. (1962) Receptive fields,
binocular interaction and functional architecture in the cat's
visual cortex. *J. Physiol., Lond. 160* 106-154.

Huffman, D. A. (1971) Impossible objects as nonsense
sentences. In *Machine Intelligence 6,* Eds. R. Meltzer & D.
Michie, pp 295-323. Edinburgh: The Edinburgh University Press.

Ittelson, W. H. (1960) *Visual space perception,* pp145-147.
New York: Springer.

Julesz, B. (1971) *Foundations of Cyclopean perception.*
Chicago: The University of Chicago Press.

Julesz, B. (1975) Experiments in the visual perception of
texture. *Scientific American 232,* 34-43, April.

Julesz, B. & Miller, J. E. (1976) Independent spatial-

frequency-tuned channels in binocular fusion and rivalry.
Perception 4, 125-143.

Kolers, P. A. (1972) *Aspects of motion perception.* New York:
Pergamon Press.

Land, E. H. & McCann, J. J. (1971) Lightness and retinex
theory. *J. opt. Soc. Am. 61,* 1-11.

Mackworth, A. K. (1973) Interpreting pictures of polyhedral
scenes. *Artificial Intelligence 4,* 121-138.

Marr, D. (1974) A note on the computation of binocular
disparity in a symbolic, low-level visual processor. *M.I.T.
A.I. Lab. Memo 327.*

Marr, D. (1976) Early processing of visual information. *Phil.
Trans. Roy. Soc. B. 275,* 483-524.

Marr, D. (1977) Analysis of occluding contour. *Proc. Roy.
Soc. B. 197,* 441-475.

Marr,D. (1977b) Artificial Intelligence - a personal view.
Artificial Intelligence 9, 37-48.

Marr, D. & Nishihara, H. K. (1977) Representation and
recognition of the spatial organization of three-dimensional
shapes. *Proc. Roy. Soc. B. 200,* 269-294.

Marr, D. & Poggio, T. (1976) Cooperative computation of
stereo disparity. *Science 194,* 283-287.

Marr, D. & Poggio, T.   (1977a) From understanding computation to understanding neural circuitry. *Neurosciences Res. Prog. Bull. 15,* 470-488.

Marr, D.   & Poggio, T.   (1977b) A theory of human stereo vision. *M.I.T. A.I. Lab. Memo 451.*

Marr, D. , Poggio, T. & Palm, G.   (1977c) Analysis of a cooperative stereo algorithm. *Biol. Cybernetics 28,* 223-239.

Mayhew, J. E. W. & Frisby, J. P.   (1976) Rivalrous texture stereograms. *Nature 264,* 53-56.

Nevatia, R.   (1974) Structured descriptions of complex curved objects for recognition and visual memory. *Stanford A.I. Memo 250.*

O'Callaghan, J. F.   (1974) Computing the perceptual boundaries of dot patterns. *Computer Graphics and Image Processing 3,* 141-162.

Richards, W. A.   (1971) Anomalous stereoscopic depth perception. *J. opt. Soc Amer. 61,* 410-414.

Richards, W. A.   (1977) Stereopsis with and without monocular cues. *Vision Res. 17,* 967-969.

Riley, M.   (1977) Discriminant functions in early visual processing. (In preparation).

Rosenfeld, A. , Hummel, R. A. & Zucker, S. W.   Scene labelling by relaxation operations. *IEEE Transactions on Systems, Man and Cybernetics, SMC-6,* 420-433.

Schatz, B. R.   (1977) On the computation of texture discrimination. *To be given at the Fifth International Joint Conference on Artificial Intelligence, August 1977.*

Stevens, K. A.   (1977) Computation of locally parallel structure. *Biol. Cybernetics 29,* 19-28.

Tenenbaum, J. M. & Barrow, H. G.   (1976) Experiments in interpretation-guided segmentation. *Stanford Research Institute Technical Note 123.*

Ullman, S.   (1976a) On visual detection of light sources. *Biol. Cybernetics 21,* 205-212.

Ullman, S.   (1976b) Filling-in the gaps: The shape of subjective contours and a model for their generation. *Biol. Cybernetics 25,* 1-6.

Ullman, S.   (1977) The interpretation of visual motion. *M.I.T. Ph. D. Thesis, June.*

Wallach, H.   & O'Connell, D. N.   (1953) The kinetic depth effect. *J. exp. Psychol. 46*, 205-217.

Waltz, D.   (1975) Understanding line drawings of scenes with shadows. In: *The psychology of computer vision*, Ed. P. H. Winston, pp19-91. New York: McGraw-Hill.

Lectures on Mathematics in the Life Sciences
Volume 10, 1978

CONTROL OF MITOSIS AND TISSUE GROWTH IN THREE DIMENSIONS

Ronald M. Shymko[1]

ABSTRACT. A growth control mechanism is considered in which a self-regulatory factor, possibly a diffusible inhibitor, acts at a localized "transition point" in the mitotic cycle, while nutrients have more generalized effects, uniformly regulating cellular processes and rate of passage through the cycle. It is shown that geometry plays a major role in determining the patterns of growth and cell death, by limiting the distribution of nutrients and other regulatory factors. These ideas are applied to growth in a spheroidal geometry where nutrient limitation has important effects, and it is shown how the effects of nutrient and cell-derived control factors might be discriminated. Analysis of experimental results on growth in spheroidal tissue culture indicates that the observed growth regulation is not due to nutrient alone, but that some other control mechanism is also acting. An assessment of possible factors affecting growth, including toxic products accumulating in the spheroid core, suggests that a diffusible inhibitor secreted by the cells may be acting to limit cell number, while overall growth rate is primarily determined by nutrient. Predictions made on this basis are shown to agree quantitatively with experiment.

AMS(MOS) subject classifications (1970). Primary 92A05; secondary 93C15, 93C50, 93D15.

[1]Research partially supported by the National Institutes of Health under Grant 5T 01-AM-05150-18.

1. INTRODUCTION. Growth rate in a cell population is determined by the rate of passage of the individual cells through
the mitotic cycle. Progress through the cycle can be affected by
regulatory factors derived from the cells themselves, or by factors present in the external environment, particularly nutrients

A control factor might act, for example, at a specific point i
the mitotic cycle, gating cells through this point in their cycle
but leaving the rest of the cycle unaffected [1,2]. There is evidence that some internal growth regulators act in this way. Mutant
have been isolated in yeast which arrest cells at particlar points
in the cycle [3], suggesting the existence of "transition points"
[2] in the cycle under the control of specific gene products. A com
mon observation in cells under density dependent inhibition is tha
these cells are arrested in their cell cycle somewhere in G_1, the
period between mitosis and nuclear DNA replication [4]. Furthermore
those cells escaping the block appear to pass through the remainder
of the cycle at a near normal rate [5]. This suggests that a transition point in G_1 is a site of self-regulation of growth [6].

Variations in nutrient can strongly affect the cell cycle.
It is known that decreasing the serum level in tissue culture
medium can increase the interdivision interval in some cells to
several times its minimum value [7]. Although any given nutrien
may act in only a short segment of the cycle, there is evidence
that many events in the cycle can be affected to some extent by
changes in total nutrient. For example, nutrient deprivation
often affects passage through G_1 [8], but can also strongly affect the remainder of the cycle, as demonstrated in organisms
where the G_1 phase is entirely lacking [9]. In addition, since
all cellular processes require energy, the average rate at which
these processes occur should depend on the cell's nutrient suppl
I will consider the simplest case, in which all processes are
affected uniformly by moderate changes in the rate at which a
cell receives nutrient.

Whatever the exact effects of nutrient are, it will be true that cells with greater access to nutrient will have a proliferative advantage over cells receiving less nutrient. This means that those cells whose surfaces are nearest the nutrient source can be expected to grow faster than other cells remote from the source. Similarly, the average growth rate in regions of high cell density should be lower than in regions of low density, due to competition between cells for a limited nutrient supply. Therefore, growth control due to nutrient will depend on the spatial arrangement of cells, that is, on the geometry of growth.

Control of growth due to cell-derived factors can also be strongly affected by geometry. Normal cells growing in tissue culture divide rapidly at low cell density but slow their division rate and often virtually stop dividing when a critical cell density is reached [10]. If this "density dependent inhibition" [11] of growth is due to internal factors, some form of communication between cells which is inhibitory to growth is implied. Communication could be achieved through direct cell-cell contact [12] or through some growth inhibitory substance which passes from cell to cell [13,14]. In either case, communication will be more efficient, and inhibition will be stronger, in regions of high cell density, or in compact cell masses, than in more dispersed cell configurations. Therefore, geometry will be critically important in determining the rate of growth.

It can be seen from the above discussion that the effects of changes of geometry on growth rate are similar whether control is due to nutrient or to a direct inhibitory interaction between cells. This makes it difficult, in many cases, to distinguish which of these factors might be responsible for an observed phenomenon of growth, even so basic a phenomenon as density-dependent inhibition [10]. The purpose of this article is to discuss the consequences of growth control by both nutrient and internal inhibitory factors, and to show how the distinction between their effects might be made. In order to show the effects of geometry most clearly, I will assume that the inhibitory cell-cell interaction is due to a diffusible inhibitor secreted by the cells. Most of the arguments will be qualitative and will apply equally to inhibition mediated by direct cell contact. I will consider a mechanism in which an inhibitor acts at a localized transition point in the cycle while nutrient has a generalized effect on growth, and will take into account the effects of cell death. The ideas presented will be applied to the growth of cells in spheroidal geometries, where experiments show some unexpected behavior.

2. THE CELL CYCLE. In the usual view, the cell cycle is divided into four major stages, M, G_1, S, and G_2 [15] following one another in a sequence which repeats itself after a time τ (Fig. 1). Control of growth, then, could be accom-

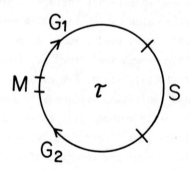

FIGURE 1. The four major stages of the cell cycle. M: mitosis, S:DNA synthesis, G_1:postmitotic gap, G_2:premitotic gap.

plished by changing the average value of τ in the cell population. This simple picture appears to be inadequate to explain certain commonly observed properties of growth in tissue culture. If the cycle is as shown in Fig. 1, and control is homogeneous over the entire cell population, the straightforward expectation is that the distribution of cell cycle times in the population

should be symmetrical about some average time τ. The actual
observed distribution in many types of cultured cells is highly
asymmetrical, with a disproportionate number of cells having
very long cycle times [16]. Secondly most of the variation in
cycle time seems to come from variation in the length of G_1,
while the remainder of the cycle is relatively fixed in length
[17]. This suggests that G_1 is the sensitive period for regulatory influences, and should be considered separately from the
rest of the cycle.

Smith and Martin [18] have analyzed the distribution of cell
cycle times for different cell types. They showed that if all
cells in the population have exactly the same cycle time, the
fraction α of cells of a certain age remaining in interphase is
100% until their age equals the cycle time and is 0% thereafter
(Fig. 2A). With a Gaussian distribution in cycle times about

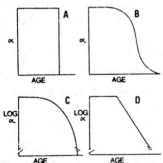

Reprinted from J. A. Smith and L. Martin, *Do cells cycle?*, Proc. Nat. Acad. Sci.
U.S.A., vol. 70 (1973) p. 1263. See [18].

FIGURE 2. The proportion of cells remaining in interphase
as a function of age. Explanation in text.

some average time, the sharp edge in the curve becomes rounded (Fig. 2B),
which translates into a quadratically-decreasing curve when replotted
as log α versus age (Fig. 2C). The experimentally observed distribu-
tions, as plotted by Smith and Martin from data on several cell types,
are shown in Fig. 3. The curve remains roughly flat until some critical age, different for different cell types, and drops off <u>linearly</u>

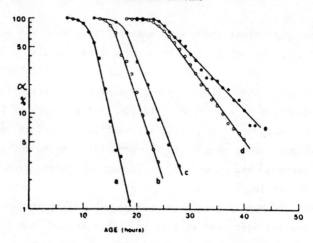

Reprinted from J. A. Smith and L. Martin, *Do cells cycle?*, Proc. Nat. Acad. S(
U.S.A., vol. 70 (1973) p. 1263. See [18].

FIGURE 3.   Distribution of generation times of various cell
types in culture.   Curves have slope -P and "shoulder" at age T_B
(a) rat sarcoma; $P(hr^{-1})=0.45$, $T_B(hr)=9.5\pm1.0$ (b) HeLa S3; $P=0.32$
$T_B=14\pm0.8$ (c) mouse fibroblasts; $P=0.30$, $T_B=15\pm1.2$ (d) L5 cells
$P=0.18$, $T_B=22.5\pm1.4$ (e) HeLa; $P=0.14$, $T_B=23\pm0.8$.

from that point, with slope also dependent on the cell type.   To
explain this, Smith and Martin proposed a probabilistic model,
similar to others proposed in earlier work [19,20].   A simplified
version is shown in Fig. 4.

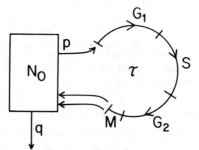

FIGURE 4.   Probabilistic model of the cell cycle. N_0 cells
in a "resting state" have probability p of entering the mitotic
pathway and probability q of dying.

Soon after mitosis, cells enter a resting state   (Smith and

Martin's "A-state"), from which they have probability p per unit time of reentering the mitotic pathway. The duration of the mitotic pathway remains constant under constant external conditions. Growth is balanced by cell death, which I have assumed takes place with probability q per unit time from the resting state. If the rate of cell death is low, the plot of log α versus age has the form shown in Fig. 2D. The "shoulder" of the curve occurs at age τ and the slope of the linear region is -p. The experimental results of Fig. 3, therefore, can be predicted with different values of p and τ for the different cell types.

Defining the number of cells in the resting state as N_0 and the number in the mitotic pathway as N_m, growth of the cell population is given by a pair of differential equations

$$dN_0/dt = -pN_0 - qN_0 + 2N_m/\tau$$

$$dN_m/dt = pN_0 - N_m/\tau.$$

(1)

I have assumed, for simplicity, that the cells, N_m, in the mitotic pathway are uniformly distributed around that sequence, so that the rate of cell division is proportional to N_m. A better approximation would take into account the time delay τ, between a cell's entry into and exit from the mitotic pathway [20]. The qualitative results are not changed by the simplification in Eqs. 1.

A cell's average intermitotic time, disregarding cell death, is the sum of the time τ spent in the mitotic pathway and the average time spent in the resting state. If there are initially N cells in the resting state which will eventually divide, Ne^{-pt} of them will remain at time t. Then, the number of cells which leave between times t and t + dt is $Npe^{-pt}dt$. Therefore, the average time spent in the resting state is given by

$$(1/N)\int_0^\infty Npte^{-pt}dt = 1/p.$$

(2)

All cells which divide spend a time τ in the mitotic pathway, so that the average cell cycle time among the population is

$$T_{av} = \tau + 1/p. \tag{3}$$

3. CONTROL IN THE CELL CYCLE. A change in either τ or p changes the total cell cycle time. Given the body of evidence that control in tissue culture very often seems to occur in "G_1" (that is, between mitosis and the beginning of nuclear DNA replication) [4], I will assume that the action of cell-derived inhibition is on the parameter p . I will further assume that nutrient has a generalized effect on the cycle time, affecting the rate of both entry into and passage through the mitotic pathway. This can be emphasized by redefining the parameters p and q as the probabilities per time τ of entering the mitotic pathway or of dying. Then Eqs. 1 become

$$dN_o/dt = (-pN_o - qN_o + 2N_m)/\tau$$

$$dN_o/dt = (pN_o - N_m)/\tau. \tag{4}$$

Cell number reaches a steady state when p=q. At the steady state $N_m = pN_o = pN/(1+p)$ where N is the total viable cell number, so that the rate at which new cells appear and old cells die is $N_m/\tau = pN/(1+p)\tau$.

In the simplest case, nutrient determines τ and inhibitor determines p. I will make the important assumption that the rate of all cellular processes, and hence $1/\tau$, is proportional to the rate at which a cell receives nutrient. The death rate, q , depends on the exact conditions of growth, but an interesting possibility is that q is constant over some range of nutrient level [21]. In this case, since the probability of death is constant per time interval τ, a cell's average lifetime is a constant number of cycles, rather than a constant time. There is some evidence for this kind of behavior from experiments on cellular senescence in cultured human cells [22]. Of course, below some critical level of nutrient all cells will die, so that q must increase rapidly as this level is approached.

If nutrient affects only τ in Eqs. (4), it simply scales the timing of events. For example, the average cell cycle time from Eq. (3) is now given by

$$T_{av} = \tau(1 + 1/p). \qquad (5)$$

Control on p and q can occur independently, and growth patterns can be similar, although expanded or contracted in time, under widely varying nutrient conditions. Furthermore, if the cell population reaches a size at which division rate equals death rate, and cell number stops increasing, nutrient will not affect further growth, since time-scaling does not alter a net zero growth rate. This provides for highly flexible, yet stable control of growth, since growth rate can automatically adjust itself to changes in nutrient while internal control is unaffected. One possible consequence, for example, is that a stable size can be reached which is independent of the time required to reach this size.

4. EFFECTS OF GEOMETRY - GROWTH OF A SPHEROID OF CELLS. The pattern of growth in a cell population depends strongly on the distribution of nutrients and other growth regulators among the population. Geometry is a major factor affecting this distribution [23,24,25]. In aggregates of cells with no vascularization, such as in tissue culture, nutrient diffuses inward from the outer surface of the aggregate, and cells in the interior obtain less total nutrient than cells on the surface. For example, oxygen is an important nutrient which can diffuse only 100-200 μm into growing tissue [26]. Therefore, since a cell's growth rate depends on the nutrient it receives, the average growth rate in a population of cells will increase with an increasing proportion of cells on the surface of the aggregate, where they are in direct contact with the nutrient supply. A simple geometry where this

occurs is the spheroid.

One prediction about growth in a spheroid that can be made on the basis of nutrient alone is that, as the spheroid grows, cells will eventually begin dying in the center due to lack of nutrient. If oxygen is the limiting nutrient, for example, a necrotic core should appear when the spheroid is 200-400 μm in diameter. If no other control factor is operating, the viable cells should be confined to a surface layer of approximately constant thickness, $d \simeq 100\text{-}200$ μm, as growth proceeds. The volume of this surface layer is then approximately equal to Sd, where S is the surface area of the spheroid. If the average growth rate in the layer is constant over time, and if there is no loss of volume from the necrotic core, the rate of volume increase in the spheroid will be proportional to the surface area, or the square of the radius, R. Therefore, $R^2 dR/dt \propto R^2$ which implies that

$$R = k_1 t. \tag{6}$$

i.e. the radius, and hence the diameter, increases <u>linearly</u> in time. This result has been discussed by others [27,28].

A further prediction can be made by noting that if the average cell volume among the population remains constant as the spheroid expands, the number of viable cells will be proportional to the volume of the surface layer of constant thickness d. The volume of this layer is proportional to the surface area, or to R^2, so that the prediction is that viable cell number will increase quadratically in time,

$$N = k_2 t^2, \tag{7}$$

as diameter increases linearly.

The above predictions were made assuming that nutrient is the only control factor acting in the spheroid. Experiments show that nutrient alone is not sufficient to account for the observed growth patterns.

5. EXPERIMENTAL RESULTS ON SPHEROIDAL TISSUE CULTURE. Very
valuable data on growth of cells in spheroidal tissue culture
comes from the work of Folkman and Hochberg [29]. Cells were
suspended in soft nutrient agar, where they began to grow, and
the growing spheroid was transferred to fresh nutrient agar every
2-3 days. Different cell types followed the same general growth
pattern. After an initial period of rapid growth, a necrotic
core appeared in the center of the spheroid when the spheroid was
less than 400 μm in diameter. This necrotic core increased in
size as the spheroid grew, and the proliferating cells were confined to a surface layer. Between the proliferating cells and
the necrotic core was a layer of nonproliferating but viable
cells. The size of the spheroids increased steadily until,
finally, a maximum diameter was reached which remained constant
through repeated transfers. At this diameter, the proliferating
cells formed a layer only 1-2 cells deep on the surface. The
reason for this growth limitation, as suggested by Folkman and
Hochberg, is that there is loss of volume from the necrotic core
as cells lyse and their contents diffuse outward through the surface of the spheroid. This loss apparently increases until, at
the dormant diameter, growth in the thin layer of proliferating
cells balances the volume loss from the core.

Folkman and Hochberg measured the increase in the diameter
of a growing spheroid by projecting its shadow onto a flat surface. The pattern of increase for V-79 Chinese hamster lung
cells is shown in Fig. 5. Note that the growth is roughly linear
over the entire growth period. If volume loss from the necrotic
core is small until very late in growth, this growth curve is
therefore consistent with nutrient being the only important
growth factor acting in the spheroid.

In parallel experiments, Folkman and Hochberg dispersed spheroids by trypsinization or by simple shaking, and obtained counts
of total cell number and viable cell number as a function of

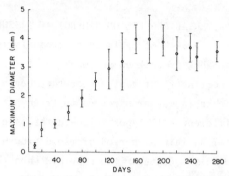

Reprinted from J. Folkman and M. Hochberg, *Self-regulation of growth in three dimensions*, Journal of Experimental Medicine vol. 138 (1973) p. 745. See [29]

FIGURE 5. Mean diameter and standard deviation of 70 isolated spheroids of V-79 cells. Very old spheroids occasionally shattered and were discarded. Therefore, mean diameter after 200 days represented approximately 30 colonies.

age. For different cell types, they found that, instead of increasing steadily until the dormant size was reached, the cell count increased rapidly at first but peaked at a steady state value long before the spheroid stopped growing! As shown in Fig. 6 for V-79 cells, cell number saturated at about 45 days,

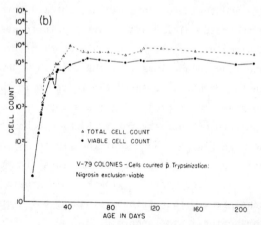

Reprinted from J. Folkman and M. Hochberg, *Self-regulation of growth in three dimensions*, Journal of Experimental Medicine vol. 138 (1973) p. 745. See [29]

FIGURE 6. The number of total cells and the number of viable cells in V-79 spheroids reach a steady-state 130 days before the spheroid stops expanding.

whereas the dormant diameter was not reached until after 175 days of growth. Furthermore, the linear increase in diameter was affected very little, or not at all, when the cell number stopped increasing. Recall that the prediction made on the basis of nutrient was that cell number would be proportional to surface area throughout growth. Since surface area rises by about a factor of 15 during the period between 45 and 175 days of growth, this prediction is not consistent with the constant cell number observed in this period.

The surprising nature of these results is emphasized in the plot of Fig. 7. This plot is based entirely on experimental

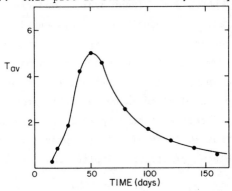

FIGURE 7. Average interdivision interval, in days, among V-79 cells growing in a spheroid. Calculated for cells of volume 2000 μm^3, from Fig. 6 and a best linear fit to Fig. 5, assuming spheroid growth is due only to cell division, with no loss from the necrotic core.

results, assuming that all volume increase is due to cell division and neglecting volume loss from the necrotic core. From the best linear fit to the graph of Fig. 5, I calculated the rate of volume increase at each time point and divided this by the corresponding viable cell number from Fig. 6 to obtain the volume increase per unit time per cell. Average division rate in the cells was then obtained by dividing this number by the average cell volume, which I estimated to be about 2000 μm^3, using data from autoradiographs of sectioned spheroids. The inverse of the divi-

The calculated cycle time rises from less than 1/2 day to 5
days after 50 days of growth (when cell number stabilizes) and
drops thereafter, reaching a low of about 16 hours when the dormant size is reached. This means that in the linear growth period between 50 and 175 days, the cells divide at an ever-increas-
ing average rate, while death rate rises along with the division
rate, keeping the cell number constant. This unexpected result
suggests that some control factor in addition to nutrient is acting in the spheroidal culture.

6. FACTORS AFFECTING GROWTH IN THE SPHEROID. The fact that
growth in the spheroid is linear over a long time period suggests
that nutrient may be responsible for this effect even though it
does not account for the cell number data. In a rough approximation, the maximum rate at which nutrient can enter through the
surface of the spheroid is proportional to the surface area, S.
If this nutrient is divided among N cells, the average nutrient
received by each cell will be proportional to S/N. In Eqs. (4),
if we assume that the rate of cellular processes is proportional
to the available nutrient, this implies that

$$1/\tau \propto S/N. \qquad (8)$$

Therefore, from Eq. (5) we see that, if p does not vary over
too large a range, the average division rate will be approximately proportional to S/N, and the total division rate will be
proportional to S. This accounts for a linear increase in diameter. Note that the increase in diameter can be nearly linear
over large changes in cell number, since the rate of total volume increase depends mainly on the total nutrient input. If cell
number remains constant, the division rate (and the death rate)
would be proportional to the increasing surface area, in agreement with the results shown in Fig. 7.

I will show in the next section that the value of p can
remain constant after a steady state cell number is reached, if
sion rate is the average cell cycle time, plotted in Fig. 7.

it is under the control of a diffusible inhibitor. To understand
the behavior of the death rate, q, several factors must be considered.

One cause of death is lack of nutrient in cells in the interior of the spheroid. If nutrient penetrates a constant distance d into the spheroid, the inner boundary of penetration
will move outward at the same velocity, dR/dt, as the outer diameter, about 11 μm per day for V-79 cells, and cells passed by
this boundary will die. The average cell density in the surface
layer is N/Sd, so that the average death rate due to the advanc-
ing penetration boundary would be approximately S(dR/dt)/Sd.
Autoradiographs of sectioned spheroids show that the necrotic
boundary remains at about a depth of 150 μm from the surface.
With this value of d, the predicted death rate is about 0.07 da^{-1},
which is much smaller than the rates, between 0.2 da^{-1} and 1.5
da^{-1} (Fig. 7) required to maintain constant cell number during
the latter stages of spheroid growth. Other causes of cell death
must therefore be considered.

Since a large necrotic core is present in the spheroid, it
is probable that toxic substances released in the core contribute
to the death rate. However, it is unlikely on general grounds
that such toxic substances can fully account for the observed
growth patterns. For, suppose that no control factors other than
nutrient and toxic products were present in the spheroid. Nutrient might account for the increasing division rate seen in the
late stages of growth as described above and as seen from the experimental data (Fig. 7). In order to account for the saturation
of cell number (Fig. 6), death rate would have to rise faster
than the division rate, due to the accumulation of toxic products,
for example, so that it can eventually overtake the division rate.
But, once this happens, there is no obvious reason why the death
rate should not continue increasing above the division rate,
since both the spheroid and its necrotic core continue growing.

Therefore, cell number would decrease after this point instead of remaining constant as observed. An additional control factor seems to be required to maintain a constant cell number.

The experimental results suggest that the effect of toxic factors is highly specific. Since death rate increases even as division rate increases, death seems to occur without affecting division rate. This might happen, for example, if a toxic factor acted at a particular point in the cell cycle, influencing a "decision" between death and further division once every cycle without otherwise inhibiting passage through the cycle. In this picture, death rate would tend to be proportional to division rate as division rate varies, for example due to variations in nutrient. In V-79 spheroids, the calculated death rate late in growth remains proportional (i.e., equal) to the division rate as division rate varies by more than a factor of 7. That is, the average cell lifetime is exactly one cycle, over large changes in cycle time. The similar "cycle-counting" behavior seen in cell senescence experiments [22] suggests the interesting possibility that a general response of cells to toxic influences is to die at a rate, determined by these influences, which is constant per average cycle time, while cycle time varies due to other environmental factors, including nutrient level.

In Eqs. (4), I have assumed that $1/\tau$ is proportional to the nutrient level and hence to S/N (Eq. (8)). Therefore, if N is constant, p must be equal to q, and both must be constant in order that total division rate and death rate be proportional to the surface area of the spheroid (that is, in order that spheroid diameter increase linearly at constant N). By Eq. (5), the average cell cycle time will be proportional to τ, which means that death rate per cycle time is constant. In the absence of further information, I will assume that q is constant throughout growth.

7. CONTROL BY A DIFFUSIBLE INHIBITOR. In the system described by Eqs. (4), the probability p of entering the mitotic
pathway is assumed to depend on the concentration of a diffusible
inhibitor secreted by the living cells. As cell number N increases in the early stages of growth, the concentration of inhibitor, ψ, in the spheroid can be expected to rise, since the
rate of production of inhibitor is proportional to N, whereas
the rate of loss is proportional to the surface area, or roughly
to $N^{2/3}$ [25]. As ψ increases, p falls until it equals q
and the net cell production rate is zero. A steady state will be
maintained only if p remains constant and equal to q. However,
p not only depends on N, through the production of inhibitor,
but it also depends on the size of the spheroid, since the inhibitor is diffusible. When a constant cell number is reached, the
spheroid continues growing as dead cells accumulate in the interior. With the assumption that the rate of all cellular processes,
including secretion of inhibitor, is proportional to nutrient,
inhibitor should be produced at a rate proportional to the surface area. In this case, the increasing production rate as the
spheroid grows will just balance the increasing loss rate, and
the concentration ψ will remain constant. That is, geometrical
factors exactly cancel out, so that ψ and hence p depend only
on N, the cell number. This means that p will decrease until
a critical cell number is reached when p = q and will remain
constant thereafter, maintaining a stable cell number as the
spheroid continues growing. Note that if q does not remain
constant after p becomes equal to q, but rises, for example,
the equilibrium cell number will drift downward. Therefore, it
is important that q be constant in Eqs. (4) in order that cell
number remain constant over long periods of time.

In summary, three major assumptions have been made: 1. that
p is controlled by a diffusible inhibitor secreted by the cells;
2. that the rate of all cellular processes is proportional to the

nutrient level; and 3. that the death rate per cycle, q, is
constant at least in the later stages of spheroid growth. With
these assumptions, the model described by Eqs. (4) predicts that
cell number in a spheroid will rise to a constant level under the
control of p and q, while the average cell cycle time, or division rate, among the cell population varies according to the
nutrient level. The total nutrient input is divided among the
cells, so that once a constant cell number is reached the average
division rate in the cells will increase in proportion to the sur
face area, and the diameter will increase linearly in time. In
the next section I will discuss the quantitative predictions of
the model.

   8. QUANTITATIVE PREDICTIONS. For V-79 cells, the minimum
cycle time when nutrient is plentiful is about 8 hr [30]. As
shown in Fig. 4, the interdivision time is minimized in a cell
which spends essentially no time in the resting state, but quickly reenters the mitotic pathway after division. Therefore, the
minimum cell cycle time is τ, the time spent in the mitotic
pathway. Nutrient controls division rate in the spheroid through
the relation (8), which can be rewritten as

$$1/\tau = kS/N. \qquad (9)$$

The proportionality constant k can be estimated by assuming
that τ will reach its lowest possible value in a single cell
when more than some critical fraction of the cell's surface area
is in contact with the nutrient medium. With a value of one-
third for this critical fraction, k is calculated from Eq. (9)
by setting $\tau = 0.33$ da, $N = 1$, and $S = 2.6 \times 10^{-4}$ mm^2, which is
one-third the surface area of a spherical cell of volume 2000 μm^3
The resulting value of k, 1.2×10^4 mm^{-2}da^{-1}, will be used in
Eq. (9) to calculate τ in Eqs. (4).

   When the spheroid is very small, the probability p of
leaving the resting state will be at its maximum value, corres-

ponding to minimal inhibition. The maximum value for V-79 cells
is not known, but a tentative estimate can be obtained by extra-
polating from the data in Fig. 3. The cell types shown have minimum values of T_B (= τ in Eqs. (4)) ranging from 9.5 hr to 23 hr,
and corresponding values of P (= p/τ in Eqs. (4)) ranging from
0.45 hr^{-1} to 0.14 hr^{-1}. If this inverse correlation holds for
V-79, these cells, with a minimum τ of 8 hr should have a maximum P of about 0.5 hr^{-1}. Then in Eqs. (4) p will be approximately equal to 4 when the spheroid is small. The average cell
cycle time among the cell population is given by Eq. (5), so that
in small spheroids the prediction is that $T_{av} \simeq 10$ hr, which ag-
rees well with the experimental values shown in Fig. 7.

     The ultimate value of p and the death rate, q, must be
estimated from the spheroid data. When the spheroid is near its
dormant diameter, the average cell cycle time T_{av} in the cells
is about 16 hr as seen from Fig. 7. At this time, the proliferating cell layer is only about 1-2 cells thick on the surface, so
that each cell should have an adequate nutrient supply. Therefore, at the dormant diameter, τ might be expected to be near
its minimum, about 8 hr, and from Eq. (5) p should be approximately 1. Since death rate equals division rate in the later
stages of growth, this fixes the death rate at the dormant diameter at q = 1. It is assumed in the following calculations that
q = 1 throughout growth.

     The drop in p from 4 to 1 is due, by hypothesis, to the
accumulation of a diffusible inhibitor as the cell number rises
to its steady state value. As discussed earlier, the concentration of inhibitor, and therefore p, depends mainly on the cell
number, and is nearly independent of spheroid size. I will assume in calculations that p drops linearly with N, falling
from 4 to 1 as cell number rises from 1 to its eventual steady
state value of 2×10^5 (Fig. 6).

     The volume increase in the spheroid can be calculated by

considering only the cell division rate. Cells dying in the interior remain there, so death has no direct effect on volume increase. The rate of cell division in Eqs. (4) is N_m/τ, so that
neglecting loss of volume from the necrotic core,

$$dV/dt = vN_m/\tau, \qquad (10)$$

where v is the cell volume, estimated to be 2000 μm^3.

With the above parameter estimates, the behavior of the system represented by Eqs. (4), (9) and (10), with a linear dependence of p on N, is shown in Figs. 8 and 9.

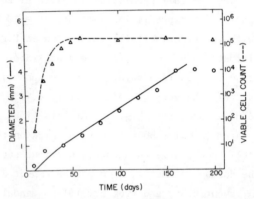

FIGURE 8.   Diameter D (left ordinate) and viable cell count
N (right ordinate) predicted from Eqs. (4), (9) and (10) with
$q = 1$; $p = 4 - 1.5 \times 10^{-5}N$ (see text); $k = 1.2 \times 10^4$ mm^{-2}da^{-1};
$S = \pi D^2$; and $v = 2 \times 10^{-6}$ mm^3. The initial condition was
$N_o = N_m = 20$ at 10 days of growth. Experimental points are plotted from Figs. 5 and 6.

Simulation was begun at an initial cell population of 40 cells
after 10 days of growth, as estimated from Fig. 6. Experimental
points are indicated for comparison. The excellent fit depends
on the value chosen for the proportionality constant k in Eq.
(9). With a different estimate for k, the time scale would be
altered, although the behavior would otherwise be the same.

Volume loss due to necrosis was not included in the calculations. It can be seen qualitatively, however, that such loss
might begin to become important at about the time that the cells

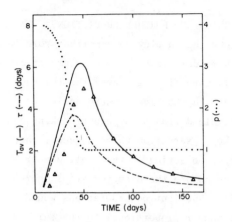

FIGURE 9. Average interdivision time, calculated from Eq. (5), in the simulation of Fig. 8. Experimental values from Fig. 7 are plotted as open triangles. Simulated behavior of p and τ is also shown.

form a layer one cell deep on the surface. At this time, the surface cell layer would begin to lose coherence - "spaces" would begin to appear between the cells. This may contribute to a relatively sudden increase in outward diffusion of necrotic products. Secondly, once the cells form a single layer on the surface, all cells will be in direct contact with the nutrient medium, so that they may all be dividing at near-maximum rates. When this happens, the average division rate in the population, which had been increasing rapidly in proportion to the surface area, will suddenly stop increasing. This may allow the loss from the necrotic core to overtake the increase due to cell division within a short time. If loss from the core is small prior to the time at which a single cell layer is reached, this could explain the observation in V-79 spheroids that the diameter stops increasing at about 4 mm, when the 2×10^5 cells form a single layer on the surface. The same correspondence between cell number and dormant diameter should be seen in other cell types.

9. DISCUSSION AND FURTHER PREDICTIONS. It is not surprising that nutrient should play an important role in the control of
growth in a spheroidal geometry. The diameter of spheroid at
which a necrotic core first appears, the thickness of the viable
cell layer, and even the linear increase in the diameter can be
predicted by simple diffusion arguments taking only nutrient into
account. The key observation which suggests that some other control factor might be acting is that cell number reaches a constant level long before the diameter stops increasing linearly
(Figs. 5 and 6). Nutrient considerations would predict an increase in cell number proportional to the surface area, which in
V-79 spheroids rises by more than a factor of 10 after a constant
cell number is attained. Therefore the result that cell number
remains constant is significant as long as the cell counts are
accurate to within a factor of 10. The observation that the proliferating cell layer narrows considerably during spheroid growth
from 5-7 cells to 1-2 cells thick, provides corroborating evidence for the cell count data. In another cell type, L-5178Y
murine leukemia, Folkman and Hochberg found that cell number actually seems to drop somewhat after reaching its peak, while the
diameter doubles. Similar observations in other cell types would
strengthen the conclusion that something in addition to nutrient
is controlling growth in the spheroid.

The plot of average cell cycle time shown in Fig. 7 was derived from Figs. 5 and 6 assuming only that all volume increase
in the spheroid is due to cell division. Since the indicated behavior is quite unexpected, alternative mechanisms for volume increase in the spheroid must be considered. A mechanism other
than cell division which might contribute to the volume increase
is an inward flow of material which bypasses the proliferating
cells entirely. Such an inward flow cannot be ruled out at this
point, but it seems unlikely that it is a dominant factor controlling spheroid growth. The diffusion of substances from the med-

ium into the spheroid would tend to increase not only in proportion to the surface area, but also in inverse proportion to the thickness of the surface cell layer. Therefore, the effect would become stronger as the spheroid grows, and cessation of spheroid growth at some dormant diameter would be difficult to achieve. Furthermore, if cell division contributes only a small fraction of the total volume increase, the average division rates among the cells would have to be much slower than those shown in Fig. 7. The calculated cycle times in Fig. 7 are between 5 days and 16 hours which are already considerably longer than the minimum interdivision time of 8 hours for V-79 cells. Cycle times much longer than this seem improbable; nevertheless, direct data on cell cycle times is needed to ultimately distinguish between the two mechanisms for volume increase.

Specific predictions can be made from Eqs. (4) about the changing proportions of the cell cycle as the spheroid grows. In the simulation of Fig. 9, p drops to the level of q in the early stages of growth and remains constant afterward. Since the relation between the number of cells in the resting state and in the mitotic pathway is given by $N_o = N_m/p$, this means that the proportion of cells in the resting state should rise initially and remain constant throughout the late period of spheroid growth. Therefore, the model predicts that "G_1" will occupy a greater proportion of the total cycle time in large spheroids than in small spheroids. This difference should appear even between very early and very late spheroids, in which the cycle times are nearly equal (Fig. 7).

The observation that cell number is constant in the late stages of spheroid growth implies that the death rate is constant per cycle time during this period. Furthermore, if cycle time behaves as shown in Fig. 7, the death rate per cycle must remain constant over large changes in cycle time. If this result is verified by direct comparison of death rates and division rates

in growing spheroids, it raises the question of how cell death
rate is controlled in the spheroid. The high degree of regulation suggests the possibility that cells themselves control their
death rate by maintaining a constant probability of death per
cycle over a moderate range of cycle times.

Two general predictions can be made about growth of different cell types in spheroids:

1. As discussed in the preceding section, the dormant diameter for any cell type should be reached near the time the cells
form a single layer on the surface of the spheroid. Therefore
the dormant diameter should be approximately the same for cell
types reaching the same maximum cell number in a spheroid. This
argument should hold as long as the effective permeability of the
spheroid surface is the same for the different cell types. If
the permeabilities are not equal, for example because one cell
type is less cohesive than the other, nutrient would be expected
to penetrate farther into the less cohesive spheroid and simultaneously, necrotic products would escape more readily. The predicted results are, first, that overall growth will be more rapid
in the less cohesive spheroid, due to an increased nutrient supply; and second, that even with more rapid growth, the dormant
diameter will be _smaller_ than in the spheroid of more cohesive
cells, since minimum cell cycle time and maximum rate of diffusive volume loss will be reached with a thicker layer of viable
cells on the spheroid surface. Folkman and Hochberg cultured
L-5178Y murine leukemia cells, which are less cohesive than V-79
cells, in a spheroid and found that the maximum viable cell count
was 3×10^5, larger than the number for V-79, whereas the dormant
diameter was about 3.8 mm, less than the 4.0 mm found for V-79.
Furthermore, the maximum cell number was reached in 20 days of
growth and the dormant diameter in 40-50 days in L-5178 spheroid
compared with the respective times of 45 days and 175 days in
V-79 spheroids. The predictions are therefore quite well con-

firmed for these cells.

2. Another prediction is based on varying the death rate q. If all cell functions remain the same but q is increased, Eqs. (4) show that the cell number increases more slowly, so that a given steady state cell number is reached after a longer period of growth. However, simulations show that the rate of growth of the spheroid diameter is increased, for example, from 0.026 mm/da for $q = 1$ to 0.032 mm/da for $q = 2$. While it is difficult to apply this prediction to a single cell type, since changing q changes the steady state value of N and thus changes the growth curves, it does suggest that among different cell types, more rapid growth of the spheroid is associated with increased cell death. This result may be connected in some way with the observation that there is typically a high degree of cell death in rapidly-growing solid tumors [14].

10. REFERENCES

1. Mazia, D. (1961). In The Cell (J. Brachet and A. E. Mirsky,, eds.) Vol. 3, pp. 77-412. Academic Press, New York and London.
2. Mitchison, J. M. (1971). The Biology of the Cell Cycle. Cambridge University Press, Cambridge.
3. Hartwell, L. (1974). Science 183, 46.
4. Epifanova, O. I. and Terskikh, V. V. (1969). Cell Tissue Kinet. 2, 75.
5. Cameron, I. L. and Greulich, R. C. (1963). J. Cell Biol. 18, 31.
6. Prescott, D. M. (1976). Adv. in Genetics 18, 99.
7. Cohen, E. P. and Eagle, H. (1961). J. Exp. Med. 113, 467.
8. Tobey, R. A., Anderson, E. C. and Petersen, D. F. (1967). J. Cell Biol. 35, 53.
9. Telatnyk, M. M. and Guttes, E. (1972). J. Cell Sci. 11, 49.
10. Stoker, M. (1967). Curr. Top. Dev. Biol. 2, 107.
11. Stoker, M. G. and Rubin, H. (1967). Nature 215, 171.
12. Abercrombie, M. and Heaysman, J. E. M. (1954). Exp. Cell. Res. 6, 293.
13. Loewenstein, W. R. and Kanno, Y. (1966). Nature 209, 1248.
14. Bullough, W. S. and Deol, J. U. R. (1971). Symp. Soc. Exper. Biol. 25, 255.
15. Howard, A. and Pelc, S. R. (1951). Exp. Cell Res. 2, 178.
16. Nachtwey, D. S. and Cameron, I. L. (1968). In Methods in

Cell Physiology (D. M. Prescott, ed.), Vol III, p. 213.
Adademic Press, New York.

17. Prescott, D. M. (1968). Cancer Res. 28, 1815.
18. Smith, J. A. and Martin, L. (1973). Proc. Nat. Acad. Sci. (USA) 70, 1263.
19. Lajtha, L. G., Oliver, R. and Gurney, C. W. (1962). Brit. J. Haemat. 8, 442.
20. Burns, F. J. and Tannock, I. F. (1970). Cell Tissue Kinet. 3, 321.
21. Littlefield, J. W. (1976). Variation, Senescence, and Neoplasia in Cultured Somatic Cells. Harvard University Press, Cambridge, Massachusetts and London.
22. Dell'Orco, R. T., Mertens, J. G. and Kruse, P. F. (1973). Exp. Cell Res. 77, 356.
23. Greenspan, H. P. (1972). Stud. Appl. Math. 51, 317.
24. Folkman, J. and Greenspan, H. P. (1975). Biochim. et Biophys. Acta 417, 211.
25. Shymko, R. M. and Glass, L. (1976). J. Theor. Biol. 63, 355
26. Thomlinson, R. H. and Gray, L. H. (1955). Brit. J. Cancer 9, 539.
27. Mayneord, W. V. (1932). Amer. J. Cancer 16, 841.
28. Laird, A. K. (1964). Brit. J. Cancer 18, 490.
29. Folkman, J. and Hochberg, M. (1973). J. Exp. Med. 138, 745.
30. Klevecz, R. R. (1976). Proc. Nat. Acad. Sci. (USA) 73, 401

DEPARTMENT OF BIOCHEMISTRY AND BIOPHYSICS
SCHOOL OF MEDICINE
UNIVERSITY OF PENNSYLVANIA
PHILADELPHIA, PENNSYLVANIA 19104

CONTROL OF SEQUENTIAL COMPARTMENT FORMATION IN <u>DROSOPHILA</u> *

Stuart A. Kauffman, Ronald M. Shymko and Kenneth Trabert

INTRODUCTION

Among the most fundamental tasks faced by the developing embryo is the reliable assignment of different developmental programs to the proper regions of the embryo. In many different organisms the fate of any region becomes progressively restricted. Recent discoveries indicate that in <u>Drosophila</u> <u>melanogaster</u> this is manifested by the formation of lines of clonal restriction, called compartmental boundaries (1-7), which arise sequentially and subdivide the egg into discrete regions in the earliest stages of embryogenesis. Later, additional lines subdivide imaginal discs, the larval anlagen that metamorphose into the different adult appendages. The sequence and geometries of these lines almost certainly reflect the order in which discrete spatial domains assume different developmental programs, and provide information about the underlying positional cues which delineate and trigger those commitments. Our purpose in this article is to discuss a simple dynamical model, that appears able to generate reliably the known sequence and geometries of compartmental boundaries in <u>Drosophila</u>, and to induce the proper developmental program in each region.

<u>Drosophila</u> <u>melanogaster</u> is a holometabolous insect. At 25°C the egg hatches about 24 hours after oviposition, the three larval instars last a total of about 4 days, pupariation and metamorphosis require about 4 days, and the adult lives several weeks. After fertilization, the zygotic nucleus undergoes 12 or 13 rapid mitoses without division of the ellipsoidal egg, thereby creating a syncytium (8). By the ninth cleavage, nuclei move outward to the cortex of the egg. After the 13th cleavage, at about 3 hours, division temporarily ceases, and cell membranes separate the cortical nuclei, creating the cellular

*Reprinted from <u>Science</u>, Vol. 199, pp. 259-270, 20 January 1978.
Copyright 1978 by the American Association for the Advancement of Science.

blastoderm. About 20 minutes later gastrulation takes place, and cell division commences.

SUMMARY

During development of _Drosophila melanogaster_, sequential commitment to alternative development programs occurs in neighboring groups of cells. These commitments appear to be reflected by lines of clonal restriction, called compartmental boundaries, which progressively subdivide the early embryo, and later the imaginal discs, which give rise to different adult appendages. We propose that a reaction-diffusion system acts throughout development and generates a sequence of differently shaped chemical patterns. These patterns account for the sequence and geometries of compartmental boundaries, and predict that each terminal compartment is specified by a unique combination of binary choices made during its formation. This binary "code" interprets coherently the patterned metaplasia seen in transdetermination and homeotic mutations.

Direct evidence shows that the nuclei of the blastoderm are still totipotent, but that these initial cells are already determined at least into anterior and posterior zones (9). Additional evidence (10) strongly suggests that, at about this time, small nests of cells in different regions of the blastoderm are set apart from the forming larval structures, and become the imaginal discs. Slightly later, each disc is "determined" to form a particular part of the adult epidermis during metamorphosis. There are the following pairs of discs: eye-antenna, labial, clypeo-labrum, humeral, first leg, second leg, third leg, wing, haltere, abdominal histoblasts and the single bilaterally fused genital disc. During larval development the discs grow to 10,000 to 40,000 cells (10). During metamorphosis, most larval structures lyse, and the ectoderm of the adult is formed by the terminal differentiation and eversion of the imaginal discs.

Using gynandromorphs to construct a fate map of the egg showing the locations of regions that will give rise to each imaginal disc (11), and with the use of gynandromorphs and mitotic recombination (10), it has been possible to establish that the initial determination processes subdivide the egg into a number of geometric domains, in each of which a group of cells related by spatial proximity, but not clonal ancestry, become determined to form a specific imaginal disc.

The discovery of the sequential formation of lines of clonal restriction rests on the genetic technique of mitotic recombination (10). A larva that is a heterozygote for a recessive gene, such as mwh (multiple wing hairs), is phenotypically normal. Irradiation of the larva during development can cause a somatic recombination event in a G_2 stage cell, leading to the formation of two daugther cells, one of which is homozygous mwh, the other homozygous for the wild-type mwh allele. The homozygous mwh cell continues to divide, generating a clone. After metamorphosis, the clone is visible as an extended mwh patch on the adult cuticle. The size, shape, and location of the clone reflect features of its history.

Operational Definition of Developmental Compartments

Mitotic recombination events resulting in marked clones can be induced by irradiation as early in development as the cellular blastoderm stage. Garcia-Bellido and co-workers (1,2) found that clones initiated at such early stages occur at arbitrary locations on the adult wing, except that they never cross a specific line on the adult surface, even though the border of such a clone might run along that line for hundreds of cells. This line divides the wing and mesothorax into anterior and posterior regions. The clones behave as though they were "respecting" the line. This line of clonal restriction provides the operational definition of a compartmental boundary separating the anterior and posterior compartments of the wing and thorax (Fig. 1).

Sequential Compartmentalization of the Early Embryo

Steiner (4), Wieshaus and Gehring (5), and Lawrence and Morata (12) have induced clones at the blastoderm stage and found that no clones cross between the adjacent prothorax, mesothorax, or metathorax structures, or between the corresponding prothoracic, mesothoracic, or metathoracic legs. Steiner (4)

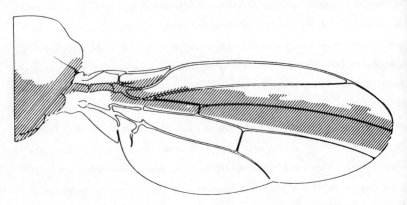

Fig. 1. A large clone whose posterior margin runs along the anterior-
        posterior border of the wing and thorax. [From Garcia-Bellido
        et al. (2)].

also found that no clone crossed between the anterior and posterior halves of

each leg. Therefore it appears that lines of clone restriction separate the

three thoracic segments, and even subdivide each of these into anterior and

posterior compartments, by the blastoderm stage. However, clones induced at

the same stage do cross from wing to mesothoracic leg, from metathorax to meta-

thoracic leg, and from left to right prothoracic leg (4,5,12).

Lawrence and Morata (12) found that clones induced a few hours after the

blastoderm stage no longer crossed from wing to mesothoracic leg. These data

demonstrate that lines of clone restriction subdivide the blastoderm longitu-

dinally prior to the establishment of restriction lines separating the dorsal

(prothorax, mesothorax, metathorax) discs from the ventral (leg) discs. It

seems clear that lines of clone restriction arise sequentially as the egg

matures past the blastoderm stage, and progressively isolate the groups of

cells that comprise the initial imaginal disc anlagen.

Sequential Compartmentalization in the Wing Disc

The anterior-posterior compartmental boundary subdivides the wing disc by

the blastoderm stage (1,2). Clones marked at later stages of development always

respect this boundary. However, Garcia-Bellido and co-workers have discovered

in the wing disc that further compartmental boundaries are formed in a specific

sequence and at specific stages of development. Among flies irradiated prior to the formation of each compartmental boundary some clones will be seen crossing that boundary; in flies irradiated after the formation of the boundary, no clones will be seen straddling that line.

The observed compartments and their times of appearance are: (i) anterior-posterior at about 3 to 10 hours; (ii) dorsal-ventral at about 30 to 40 hours; (iii) wing-thorax at about 30 to 40 hours; (iv) scutum-scutellum, and postscutum-postscutellum at about 70 hours: (v) proximal-distal wing at about 96 hours.

Data for the occurrence and timing of the later compartments in the wing disc are less secure than for the earlier compartments, both because good surface markers are sparse in some areas of the wing and thorax, and because clones marked late in development are small, making compartmental boundaries harder to observe. To overcome the latter problem, Morata and Ripoll (13) introduced and Garica-Bellido <u>et al</u>. (1,2) made use of flies heterozygotic for a mutant gene, <u>Minute</u>, M/M$^+$. <u>Minute</u> flies develop more slowly than wild-type flies. Irradiation produces some M$^+$/M$^+$ cells which grow faster than M/M$^+$ cells, thereby creating large clones which "fill up" most of a compartment and help establish it boundaries. Because of uncertainties about the relative developmental ages of <u>Minute</u> heterozygotes and wild-type flies, the timing of later compartmental boundaries is only approximate.

The five compartmental boundaries are shown schematically in Fig. 2a. In Fig. 2b they are shown projected onto the fate map of the third instar wing disc (14). During metamorphosis, the wing disc "folds" along the dorsal-ventral lines, and apposes dorsal and ventral thorax, while the wing everts.

Projecting the successive compartmental lines onto the mature third instar disc (Fig. 2b) may not accurately reflect their detailed geometries on the growing disc at the times the lines were first formed. In fact, the first anterior-posterior boundary is actually present on the blastoderm (4,5) before the wing and mesothoracic leg are clonally isolated and simultaneously divides wing and leg into anterior and posterior compartments. The remaining four boundaries form on the wing disc proper as it grows and changes shape. Each of the first three boundaries successively bisects all the compartments created

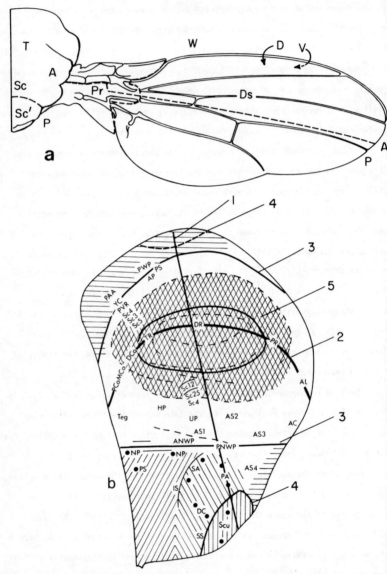

Fig. 2.   (a) The five compartmental lines on the thorax and wing. A, anterior;
         P, posterior; V, ventral; D, dorsal; T, thorax; W, wing; Pr, proximal
         wing; Ds, distal wing; Sc, scutum; Sc', scutellum. [From Garcia-Bellido
         (1)] (b) Projection of the five compartmental boundaries onto the
         fate map [Bryant (14)] of the third instar wing disc. The dotted
         line 4 is the postulated compartmental line in the ventral thorax
         needed to complete the anterior-posterior, dorsal-ventral (twofold)
         symmetry.

by the earlier lines. The last two each seem simultaneously to bisect two or more previous compartments. Garcia-Bellido et al. (2) point out the astonishing anterior-posterior and dorsal-ventral (that is, twofold) symmetries of the process. Indeed, the only deviation from complete twofold symmetry is the lack, at the fourth compartmentalization, of a line bisecting ventral thorax. But, in fact, that line may well exist, for Garcia-Bellido et al. (2) happened only to look in the dorsal thorax at that time.

If wing disc compartmental boundaries separate groups of cells with different developmental commitments, then the geometries of the lines suggest that a terminal wing disc compartment is specified by a unique combination of binary names; for example, anterior, not posterior; dorsal, not ventral; wing, not thorax; proximal, not distal.

Different Compartments Appear to Have Different Commitments

At least some lines of clone restriction almost certainly separate groups of cells with distinct developmental commitments. The different imaginal discs are in different states of determination. Determination is operationally defined as a clonally heritable state (10); therefore, determination of groups of cells to one or another disc implies clonal isolation of each group. Since neighboring domains on the blastoderm become determined to form different imaginal discs, such alternate commitments almost certainly generate the lines of clone restriction which sequentially subdivide the early embryo.

Compartmental boundaries subdividing the wing disc also seem to separate domains of cells carrying different developmental commitments. In this case, the criterion of clonal heritability cannot be applied since, during culture of disc fragments, regeneration across wing compartmental boundaries can occur (15). However, in several homeotic mutants, which convert one disc or disc region to the commitment of another disc or disc region, the boundaries of action of the mutants coincide with the boundaries of wing compartments (16,17). This suggests that a homeotic gene can have a compartment as its domain of action, and that intradisc compartments are distinctly committed cell populations (1, 7).

Some apparent lines of clone restriction may not separate domains of

distinctly committed cells. Most critically, mere clonal isolation itself can
never prove that the two isolated populations of cells are different. For example,
a line of cell death might isolate two sub-populations of identical cells.
Apparent lines of clone restriction might be artifacts because of the fusion
of initially separated regions of cells spanned by early large clones, but
not late small clones; or because late clones are small and unlikely to span
between distant structures. However, use of large M^+ clones in Minute heterozygotes obviates most of the latter problems; and it appears likely that, in
most cases, a compartmental boundary does divide a contiguous group of cells,
as hypothesized by Garcia-Bellido et al. (2) and Crick and Lawrence (7).

The data now available indicate that compartmental lines arise sequentially
throughout development from the earliest stages of embryogenesis and progressively subdivide and isolate groups of cells spatially, probably as a reflection
of the sequential commitment of neighboring groups of cells to alternative
developmental fates. Thus we could consider that a uniform mechanism might act
throughout development to account for the number, position, symmetries, and
sequence in which compartmental boundaries form. We propose a model that
appears able to do so, a model that provides an explicit measure of the similari
of developmental programs in different compartments. Since the lines on the
wing disc are the best established, we first construct our model for the wing,
then apply it to other discs and early embryo.

Characteristic Chemical Patterns

We postulate a single biochemical system of two components, with concentrations $X(r,t)$, $Y(r,t)$, at position r, time t, which are being synthesized
and destroyed at rates $F(X,Y)$ and $G(X,Y)$, and which are diffusing throughout
the tissue. The equations for this system are:

$$\frac{\partial X}{\partial t} = F(X,Y) + D_1 \nabla^2 X$$

$$\frac{\partial Y}{\partial t} = G(X,Y) + D_2 \nabla^2 Y$$

(1)

Biochemical dynamical systems such as Eqs. 1 can exhibit a variety of
behaviors. In particular, as shown long ago by Turing (18) and recently by
Gmitro and Scrivin (19) and Meinhardt (20), and reviewed by Nicolis and

Prigogine (21), such a system can have the property that its spatially uniform steady state, $X = X_0$, $Y = Y_0$ [defined by the simultaneous solution of $F(X_0,Y_0) = 0$, $G(X_0,Y_0) = 0$] is stable to all spatially distributed perturbations except those perturbations whose spatial wavelengths fall in a narrow range around some specific, characteristic wavelength, 1_0. Perturbations with wavelenghts outside this range will decay to the spatially homogeneous steady state; pertubations above and below the steady state concentrations, X_0,Y_0, with wavelengths near the natural chemical wavelength, 1_0, will grow in amplitude and create a spatially inhomogeneous pattern of chemical concentrations. If the reaction functions $F(X,Y)$, $G(X,Y)$ are linear, the amplitude of such a wave pattern grows without bound. However, with appropriate non-linear reaction functions $F(X,Y)$, $G(X,Y)$ all such perturbations will either decay or grow to a maximum amplitude without oscillations in time. Therefore, this system has the capability of establishing stable, steady state spatial patterns of concentration with spatial wavelength equal to 1_0. Given expected values of diffusion constants of about 10^{-7} to 10^{-5} cm^2/sec, such dynamical systems typically have natural chemical wavelengths on the order of 50 to 150 μm (22). This is appropriate for the size of developing imaginal discs. A biochemical kinetic system exhibiting these properties, and the mathematical analysis of Eqs. 1 is given in the appendix.

Thermal noise continuously introduces small local fluctuations to the spatially uniform steady state. These can be decomposed into a Fourier series, a weighted sum of very many spatial wavelengths. Thus, thermal noise guarantees that some very small amplitude wavelength equal to 1_0 is present. The dynamical system acts as a filter and will select and amplify that wave in time while all other wavelengths are suppressed, creating a stationary spatial pattern of wavelength 1_0. This was the heart of Turing's model (18) for the spontaneous formation of a chemical pattern from a spatially homogeneous state.

Reactions in a Growing Domain Create a Sequence of Patterns

A critical feature of compartmental boundaries is that they arise in a well-defined sequence as the imaginal disc grows in size. Our Eqs. 1 support only chemical patterns with wavelengths equal to 1_0. Let this chemical system occur in a closed bounded domain; and assume that there is no flux of reactants X and Y through the boundaries. Since diffusion occurs down spatial gradients of concentration, this assumption imposes the mathematical boundary condition that the spatial gradient of concentration at the boundaries be flat along lines perpendicular to the boundaries (23). If, for example, the chemical system is in a one-dimensional domain of length L, the no-flux boundary condition constrains the spatial patterns that appear to be superpositions of one or more sinusoidally shaped functions of the form $\cos(n\pi r/L)$, ($n = 1,2,3...,$ $0 \leq r \leq L$), which have zero spatial derivative, and consequently local maxima or minima, at the boundaries $r = 0$ and $r = L$. The nth of these patterns has wavelength $2L/n$; if this pattern is present, n half-waves of length L/n fit into the domain of length L. However, our chemical system (Eqs. 1) only allows chemical patterns of one wavelength, l_0, to grow. Therefore, the boundary conditions can be satisfied and patterns can appear only when the length of the one-dimensional domain L is an integral multiple of the half-wavelength $l_0/2$; that is, $L = nl_0/2$.

Now suppose the length L of the one-dimensional domain is less than $l_0/2$. No pattern of wavelength l_0 can "fit" into this length, hence no pattern can emerge. The chemical system stays at its spatially homogeneous steady state. Let the "tissue" length gradually increase. When $L = l_0/2$ the first chemical pattern, in which one half-wavelength fits into the legnth L, will grow, with a maximum of concentration at one boundary and a minimum at the other boundary. As the length of the tissue increases beyond $L = l_0/2$ the first pattern will no longer fit, and will decay back to the homogeneous steady state. When the tissue grows to $L = l_0$, an entire cosine pattern, maximal at both ends and minimal in the middle, or vice versa, arises. In general, at length $L = nl_0/2$, n half-wavelengths fit into the domain in tandem. Thus at a discrete succession of lengths $L = l_0/2, l_0, 3l_0/2, ...nl_0/2$ distinct sinusoidal chemical patterns emerge and decay.

Nodal Lines of Sequential Patterns Create Compartments

In direct analogy to the one-dimensional case, chemical system (Eqs. 1) will produce a sequence of differently shaped chemical patterns on complex,

growing surfaces such as the wing disc. The exact shape history of the wing
disc is not yet known in detail (24), but by mid-first instar it is a single-
cell layer backed by a thin peripodial membrane. We will therefore approximate
it as a planar shape. Since the wing disc is only joined to the larva by a
stalk, a natural assumption is that no flux of X and Y occurs across the edge
of the disc into the hemolymph.

On a growing two-dimensional domain, the shapes of the chemical patterns
that arise depend on both the size and shape of the domain. For example, on
a growing circular imaginal disc, the first three patterns that arise are shown
in Fig. 3, a to c. As we describe below, distortion from a circle to an ellipse
distorts the shapes of the chemical patterns.

The chemical patterns that form provide a means for drawing a succession
of compartmental lines across the disc. Since compartmental lines appear to
reflect the commitments of the divided groups of cells to two alternative
fates, it is natural to assume that each of the chemical patterns induces
cells in different regions of the disc to adopt one of two different commitments.
The steady state concentrations, X_0 and Y_0, occur as the nodal lines of each
chemical pattern, dividing the disc into alternate regions with concentrations
above and below the steady state (Figs. 3, a to c). Therefore we suppose X_0
is the threshold concentration of a morphogen, inducing one binary commitment
(for example, anterior) in cells above threshold, and the alternate commitment
(posterior) in cells below threshold and that this commitment is recorded by a
two-state memory switch or on-off "selector gene" (1) in each cell.

Our model postulates a succession of differently shaped chemical patterns
that grow and die away. The two-state switch recording the "anterior" or
"posterior" decision in each cell provides a memory device for the location
of the first line, which can thereafter decay. A different two-state switch
would be required for the compartment boundary induced by each successive pattern,
the first recording "anterior" or "posterior"; the second recording "dorsal"
or "ventral"; the third "wing" or "thorax". The sequential activation of each
binary switch by the successive formation of boundaries would therefore create
a binary combinatorial "code word" specifying each terminal compartment. We
discuss this below.

218   S. A. KAUFFMAN, R. M. SHYMKO, AND KENNETH TRABERT

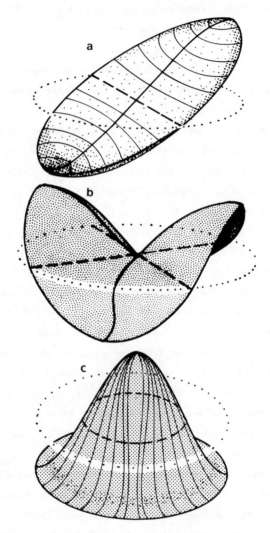

Fig. 3. (a) Wave pattern generated on a circle with scaled radius kr = 1.82.
The pattern is the product of a radial part, $J_1(kr)$ [the first order
Bessel function (26)] and an angular part, cos ϕ. The dashed nodal
line of zero (thst is, steady state) concentration runs along the
diameter of the circle from ϕ = 90° to ϕ = 270°. The dotted circle
outlines the circular radius. (b) Wave pattern, $J_2(kr)$ cos 2ϕ, generated at a scaled radius of 3.1. The dashed lines are crossed nodal
lines on two perpendicular diameters. (c) Pattern generated at a
scaled radius of 3.8, where the zero in the derivative of $J_0(kr)$
matches the radial boundary condition. The pattern is $J_0(kr)$ cos 0ϕ,
which has no angular variation. The nodal line is concentric with
the outer radius.

Each chemical pattern can arise in either of two orientations, since concentrations on one side of a nodal line could be above or below threshold. To break this symmetry, we must suppose that some inhomogeneity properly orients each chemical pattern. For example, the stalk of the disc might play such a role. An important further implication of the assumption of a single threshold X_0 is that, in a given chemical pattern, more than one nodal line might simultaneously divide the domain, creating more than one compartmental boundary, but only the two possible switch states would be induced alternately $(+,-,+,...-)$ in the neighboring regions with concentrations above and below X_0. If, in Drosophila, more than one compartmental line simultaneously arises, then more than two commitments might conceivably have been chosen. For example, the wing-thorax boundary (Fig. 2b) seems to arise as a pair of lines separating the disc into three regions. Our model assigns identical states to the two ends of the disc and the alternate state to the middle. It is interesting that, behaviorally, the two ends are thoracic and the middle is wing.

Any model of compartmentalization requires a mechanism by which clones "respect" the lines. Morata and Lawrence (17) and we propose that the alternate commitments create distinct cell affinity properties on the two sides of each line such that local cell sorting maintains the clonal integrity of compartment lines. Local sorting is consistent with the observation that marked clones remain coherent.

Chemical Patterns on an Ellipse Fit the Sequence on the Wing

As was noted earlier, the first, anterior-posterior, compartment line is formed on the blastoderm and must be explained in the context of the ellipsoidal geometry of the surface of the egg. We show below that our model accounts for this first boundary. The dorsal-ventral, wing-thorax, scutum-scutellum, and proximal-distal compartment lines arise on the wing disc itself and should closely match the lines predicted by our model on an appropriate geometry. Two factors dictate the choice of an ellipse for our approximation instead of,

for example, a circle (Fig. 3, a to c). First, throughout larval life the true
wing disc is more similar to an ellipse than to a circle (24). Second, the
ellipse is the most complex planar figure for which the succession of chemical
patterns can be investigated analytically; thereafter, detailed simulations
or other approximation techniques become necessary.

   On the circle, the nodal lines form along the radii and along the concentric circles. That is, the model compartmental lines run along lines that
are the natural coordinate system for the circle. On an ellipse, the natural
coordinate system (25) (Fig. 4) consists of concentric ellipses rather than
concentric circles, and pairs of confocal hyperbolae rather than radii. Thus,
on an ellipse, model compartmental boundaries form along these lines. As the
ellipse grows, the modes that occur are given by products of cosine- or sine-
elliptic Mathieu functions (25,26,27) of the form $Ce_n(\xi,s_{nj})ce_n(\eta,s_{nj})$ or
$Se_n(\xi,s_{nj})se_n(\eta,s_{nj})$, which we abbreviate Ce_{nj} or Se_{nj} (see appendix for explanation of the symbols). For most of the modes we consider, $j = 1$, and therefore
we use the further abbreviations Ce_n or Se_n for the above modes when $j = 1$.

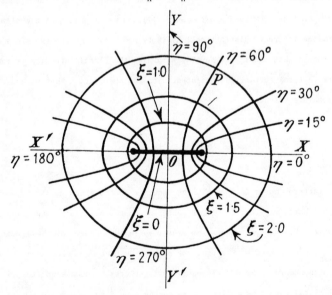

Fig. 4. The coordinates of an ellipse, ξ and η, are constant on confocal
         ellipses and hyperbolae, respectively. The interfocal distance is
         2h.

The elongation of an ellipse, relative to a circle, confers two preferred axes, the major and minor. This serves to orient successive wave patterns in constrast with the random orientation of successive patterns on the circle. The first pattern to fit on is a distortion of the first pattern on a circle, and not surprisingly for a dynamic system with a preferred wavelength, it fits on the long way first (Fig. 5A) (28). This wave pattern is a product of first cosine-elliptic functions, $Ce_1(\xi, s_{11}) ce_1(\eta, s_{11})$ or Ce_1 in our notation. Figure 5, A to F, show the succession of characteristically shaped patterns. The solid lines are the anti-symmetry nodal lines corresponding to postulated "threshold" compartmental lines. The succession of modes is Ce_1 (Fig. 5A), Se_1 (Fig. 5B), Ce_2 (Fig. 5C), Se_2 (Fig. 5D), Ce_3 (Fig. 5E), and Ce_0 (Fig. 5F).

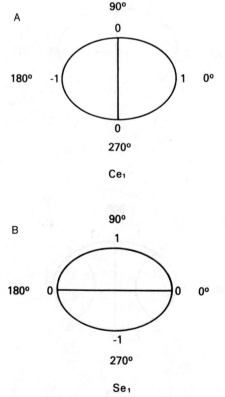

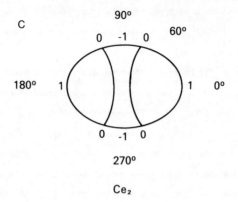

Ce_2

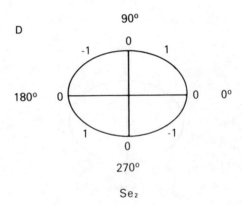

Se_2

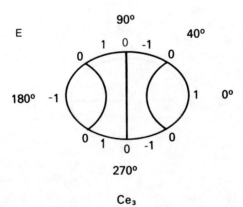

Ce_3

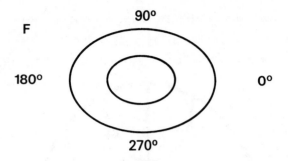

Fig. 5.   Nodal lines of successive wave patterns which fit onto an ellipse as
it enlarges. The patterns are similar to those on a circle.   (A)
$Ce_1(\xi,s_{11})(\eta,s_{11})$ (abbreviated Ce_1) is a slight distortion of
$J_1(kr) \cos \phi$ (Fig. 3a).   (B) $Se_1(\xi,s_{11})se_1(\eta,s_{11})$ (that is Se_1) is
Ce_1 rotated $90°$.   (C) Ce_2 is analogous to $J_2(kr) \cos 2\phi$ (Fig. 3b)
but on an ellipse the radii split to form pairs of confocal hyperbolae.
(D and E) Se_2 and Ce_3 are analogous to $J_2(kr) \sin 2\phi$ and $J_3(kr) \cos 3\phi$,
respectively.   (F) $Ce_0(\xi,s_{01})ce_0(\eta,s_{01})$, or Ce_0 is analogous to
$J_0(kr) \cos 0\phi$.   This mode is a hill-shaped pattern with an interior
nodal ellipse, similar to Fig. 3c.   See the appendix for definitions
of the symbols.

The first elliptical mode, Ce_1 (Fig. 5A), has a nodal line along the minor

axis of the elliptical disc which draws the first boundary on the disc proper,

the dorsal-ventral line, which also occurs the short way across the wing disc

(Fig. 2b).   The second mode, Se_1 (Fig. 5B), repeats the anterior-posterior

line.   The third mode, Ce_2 (Fig. 5C), has a pair of hyperbolic nodal lines

which nicely match the pair of lines separating ventral and dorsal thoracic

compartments from the wing compartment lying between them. Se_2 (Fig. 5D) repeats

the dorsal-ventral and anterior-posterior lines. Ce_3 (Fig. 5E) draws two nodal

hyperbolae near the dorsal and ventral narrow ends of the ellipse.   The dorsal

member of the pair draws the fourth line separating the dorsal thorax into

scutum-scutellum and postscutum-postscutellum.   The ventral member of the pair

of hyperbolae is the predicted bisection of ventral thorax needed to complete

the twofold symmetry of the observed compartment lines (Fig. 2, a and b).   The

final "hill" pattern, Ce_0 (Fig. 5F), draws a concentric interior nodal ellipse

separating proximal and distal wing.

Taking account of the growth of the wing disc, the predicted compartmental

boundaries are summarized in Fig. 6.   They compare well to the observed bound-

aries (Fig. 2B).   In short, the sequence, number, position and symmetries of

the second through fifth compartmental boundaries are almost identical to the
characteristic chemical patterns arising on a growing ellipse. The compartment
lines reflect the symmetries and natural coordinate system of an ellipse.

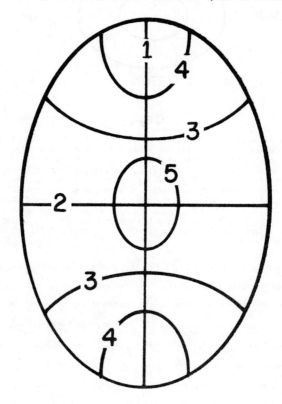

Fig. 6. Taking account of wing disc growth, we project all five predicted
compartmental boundaries onto one ellipse. The observed boundaries
are shown in Fig. 2b.

Predictions: Compartment Lines on Other Discs and the Blastoderm

Our model postulates that the chemical patterns occurring in a tissue
depend on the tissue's size and shape. It therefore makes testable predictions
about sequential compartmentalization on the different, distinctly shaped,
imaginal discs and on the ellipsoidal egg.

The compartmental boundaries that form on the leg (4), genital (6), and
haltere (11,29) as well as on the blastoderm (4,5,12) are now partially known

(Fig. 7, A to C). The occurrence of compartmental boundaries in the eye-antenna disc is in doubt (3,29).

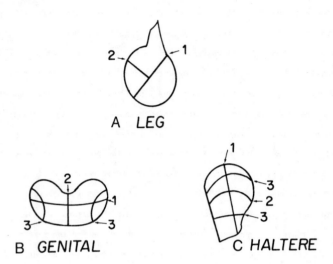

Fig. 7. Schematic compartmental lines on the (A) leg, (B) genital, and
(C) haltere.

If the unstable chemical wavelength is roughly identical in all discs, then the model predicts that the smaller of two discs with homologous shapes should form only the first few of the lines that form on the larger disc. The different imaginal discs fall into two major shape classes. Wing, haltere, and the individual leg discs are convex, fairly symmetric shapes (24,30). The eye-antenna, genital, and the pair of first leg discs, which are fused, are nonconvex, bilobed structures during most of their development (24,30). Our model makes distinct predictions for these two shape classes. As predicted, the three lines that form on the smaller haltere disc (29) are homologous to the first three that form on the larger wing disc. The first two on each leg disc match the first two on the larger wing disc. Our model cannot explain the failure of the second leg boundary to extend across the disc. On nonconvex, but symmetrical bilobed structures, one might intuitively expect that a compartmental boundary would form in the narrow isthmus between the lobes. However, none appears to form between the left and right halves of the symmetrical,

fused genital disc (6), or between the left and right members of the fused
first leg discs (4).

Our model predicts these counter-intuitive observations (see the appendix).
In symmetrical bilobed structures joined by a sufficiently narrow and short
isthmus, the chemical patterns that form must approximately fit in each lobe
separately. Thus, chemical patterns with nodal lines across the isthmus, which
would create boundaries that isolate one lobe from the other, are suppressed.
On the other hand, symmetrical patterns with maxima or minima in the isthmus
can form lengthwise along the symmetrical bilobed structure, generating the
second compartmental boundaries that arise simultaneously in each lobe of the
fused first leg discs, and in each lobe of the fused genital disc.

On the asymmetrical bilobed eye-antenna disc. Baker (3) reports that a
longitudinal compartment boundary running through the isthmus divides dorsal
eye and dorsal first antennal segment from ventral eye and the rest of the
antenna. Later, two lines form at about the same time, perpendicular to the
first line, and divide the eye into three regions. The predictions of our
model are consistent with the sequence for an appropriate asymmetric bilobed
structure. However, other workers (29) have been unable to confirm the occurrence of these compartmental boundaries in the eye-antenna disc.

The most interesting predictions of our model concern the forming blastoderm.
The egg does not grow during cleavage or later embryogenesis. However, it is
likely that, as cell membranes form and separate nuclei, resistance to diffusion
increases so that effective diffusion constants are made smaller. If diffusion
constants become smaller while the ratios of the diffusion constants do not
change, the effect is to shorten the unstable wavelength. Mathematically, the
effect is the same as if the unstable wavelength remains constant and the
spatial domain grows; as the allowed wavelength becomes shorter, a succession
of different patterns fit on the egg at a discrete succession of allowed
wavelengths.

A fundamental geometric difference between the cleaving egg and the discs
is that the egg is ellipsoidal, where as the discs are effectively planar. The
axis ratio of the egg is about 3.7 to 1 (30). As the allowed wavelength gradually
shortens, the patterns that fit on such an ellipsoid arise as longitudinal

sinusoidal patterns for the first several allowed discrete wavelengths (26-28).
These create circumferential nodal lines, which segment the egg along its
length. Only after several longitudinal segments are created is the wavelength
short enough to "fit" around the circumference of the egg. Such circumferential
patterns will create longitudinal nodal lines, and, on an ellipsoid, the first
to arise should divide the dorsal and ventral halves of the egg (Fig. 8).
Hence, the model predicts that longitudinal segmentation into at least six, or
perhaps, more, compartments should occur prior to the establishment of clonal
restrictions between dorsal and ventral discs.

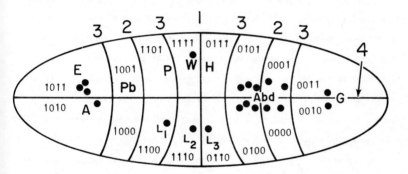

Fig. 8. Successive compartmental lines predicted on the fate map (11,49)
of the blastoderm by the chemical wave model, with the binary
combinatorial code assignment in each compartment generated by the
successive lines. A, antenna; E, eye; Pb, proboscis; Cl, clypeo-labrum;
P, prothorax; W, wing and mesothorax; H, haltere; Abd, abdominal
segments; G, genital. Assignment of clypeo-labrum to a position below
proboscis is tentative. Transdetermination predictions are given in
Table 1.

As was described earlier, the data now available confirm these predictions.
Clones formed at about 3 hours - the blastoderm stage - show restrictions to
each of the three thoracic segments, and probably also respect the anterior-
posterior compartment lines in the thoracic segments (4,5,12). However, clones
created at this time can run from mesothorax to mesothoracic leg, metathorax
to metathoracic leg, and prothorax to prothoracic leg. By 2 hours after the
blastoderm stage (12), clones no longer cross between dorsal and ventral discs.
Hence, longitudinal segmentation has in fact preceded dorsal-ventral clone
restriction, as predicted by our model.

The qualitative predictions of our model are shown in Fig. 8 for the first three longitudinal sinusoidal patterns and the first circumferential pattern. These four patterns create compartmental boundaries that would suffice to separate the major imaginal discs. A fourth longitudinal pattern, analogous to Ce_6, would simultaneously generate six circumferential compartment boundaries that would form the anterior-posterior compartment lines in the three thoracic segments, and also further subdivide the posterior and anterior thirds of the blastoderm. The thoracic portions of these predicted lines correspond to well-established boundaries. The number of circumferential compartment lines which subdivide the abdomen and the anterior third of the egg at the blastoderm stage is not yet established. The longitudinal compartmental line which separates the (dorsal) thoracic discs from their corresponding leg discs (ventral) may also extend posteriorly across the abdomen, isolating dorsal from ventral histoblast anlagen and dorsal from ventral genital anlagen (6,12). Evidence for the extension of the same "dorsal-ventral" line to the head region is in doubt (3). Our model predicts a specific sequence among the compartmental lines described above. These predictions are now being tested.

An important further prediction of the model is that the sequence and locations of compartmental lines are invariant with respect to moderate variations in the size of the egg. As the unstable wavelength becomes shorter, each successive pattern arises when the wavelength becomes short enough for that pattern to satisfy the boundary conditions. If the initial wavelength is too long to "fit" onto the egg, and eventually becomes short enough, the proper sequence of patterns will arise on the egg and be recorded by distinct binary switches. Each pattern will generate its nodal lines at proportionally identical locations on larger or smaller eggs of the same shape. Therefore, the locations of the nodal compartmental lines are size-invariant. Since there is a finite, small number of binary switches, say five, only the first five chemical patterns generate compartmental boundaries and no further pattern is recorded. Similar considerations apply to the growing imaginal disc. As long as it is initially small enough and if it eventually grows larger than required for the final recorded patttern, the locations of nodal compartmental lines will be at the proper locations. We stress that, while the nodal compartmental lines are size-invariant, they are not shape-invariant.

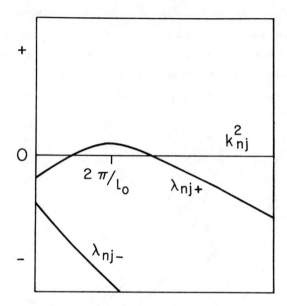

Fig. 9. Qualitative form of the two branches, λ_{nj+} and λ_{nj-}, of the dispersion relation, Eqs. A3, which gives rise to sequential pattern formation. Only those modes with wave number k_{nj} in the neighborhood of $2\pi/l_0$ will grow in time and create spatial patterns.

Certain difficulties arise in making detailed predictions about the expected sequence of compartmental lines, tissue sizes at which they occur, and their exact locations. On the wing disc, patterns Se_1 and Se_2 merely repeat previously drawn lines, and therefore the role they might play in compartment formation is not obvious. Dorsal-ventral and wing-thorax boundaries arise at about the same time, while our model predicts an interval of growth between them. On the egg, our linearized equations may predict that the anterior-posterior intrathoracic lines arise after clonal isolation of the dorsal from ventral discs. The wing disc grows 70-fold between the formation of the dorsal-ventral and proximal-distal boundaries, while our model on a perfect ellipse predicts 30-fold growth (Fig. 10). Finally the wing disc boundaries appear to arise asymmetrically in the disc. Fully nonlinear forms of Eqs. 1 generally have the properties that the expected sequence of chemical patterns can differ from the linear predictions (21,31,32); early patterns can,

for instance, suppress later patterns; and on the wing disc, for example, Ce_1
and Ce_2 might suppress Se_1 and Se_2. Also, the nodal lines might be displaced
from the locations predicted by the linearized equations (31,33). An appropriate nonlinear model, coupled with distortions from perfect ellipses and
ellipsoids, may therefore be able to circumvent many of the above difficulties.
We have limited ourselves to the restrictive predictions of the linearized
forms of Eqs. 1.

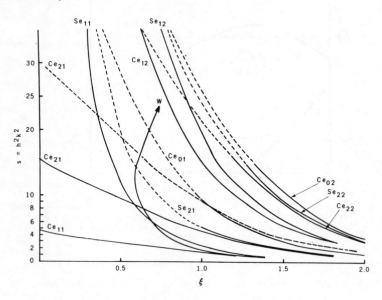

Fig. 10. Two parameters. ξ and s (see text), specify all possible ellipses.
For all ellipses whose size and shape parameters lie along one of the
curves shown, the indicated mode satisfies the no-flux boundary
condition with $k_{nj} = 2\pi/l_0$ and is therefore amplified. The dotted
portions of the curves are extrapolations of available data (28).
The arrow labeled W shows approximate shape changes during growth of
the wing disc.

Each Compartment May Be Specified by a Binary Combinatorial Code Word

The first aim of our article has been to account for the sequence and
geometries in which compartmental boundaries arise on the blastoderm and
imaginal discs. However, our model leads to an interesting hypothesis about
the form of developmental commitments in distinct compartments. It is the
nodal lines of the successive chemical patterns that appear to correspond to

the observed compartmental boundaries. The model's ability to predict the sequence and geometries of "nodal" compartmental lines is independent of how these nodal lines might be recorded. For example, a nodal line might induce special cell junctions or a line of cell death, which thereafter isolates two parts of identical cells. We stress this, since the hypothesis that compartments are distinctly committed cell populations is not conclusively established.

However, it appears highly likely that most compartmental lines do separate groups of cells as a reflection of their commitment to different developmental pathways; therefore, a natural interpretation of our model has been to postulate that the nodal lines of a pattern are a threshold level of a morphogen, which acts to induce one of two different cell choices in cells below or above threshold. By this we are led to suppose that cells in each terminal compartment have recorded the sequence of binary commitments as compartment boundaries form successively. Then each terminal compartment would be named by a unique combination of states of a small number of binary switches (Fig. 8). The possible combinations of states of the several binary switches could properly be thought of as an epigenetic code, each word of which specifies a single compartment.

The geometries of wing disc compartments also suggest that each is specified by a combination of binary names; anterior or posterior, dorsal or ventral, and so on. On the basis of these geometries and an analysis of the homeotic mutants <u>engrailed</u>, <u>bithorax</u>, and <u>post-bithorax</u>, which convert one thoracic compartment to another, Garcia-Bellido (1) and Morata and Lawrence (17) have suggested that each compartment is specified by a combination of several on-off "selector genes".

A binary code (Fig. 8) implies that each switch partitions the discs and disc compartments into two complementary subsets, those where the switch is in state 1 and all the rest, where it is in state 0. If there are genes that act only when this switch is in state 1 and others that act only when this switch is in state 0, pairs of mutants should exist that affect complementary subsets of discs. Four such pairs are already known (34,35). They provide independent evidence that each disc is specified by a combination of binary switches (35, 36), and yield a code almost identical to that predicted independently by our chemical pattern model on the blastoderm (Fig. 8) (37).

Prediction of Transdetermination and Homeotic Transformations

Two highly ordered forms of metaplasia occur in <u>Drosophila</u>: transdetermination and homeotic mutations. Transdetermination occurs when an imaginal disc determined to form one appendage is cultured in adult abdomen for several weeks, then injected into a larva for metamorphosis (10,38). Occasionally, adult cuticle typical of another disc is found in the metamorphosed implant. The patterns of transdetermination frequencies from each disc show that there are sequences of transdeterminations, and there is a global orientation toward thorax (35-38). For example, the transitions genital to antenna to wing to thorax represent one such oriented sequence. Any transdetermination altering one switch should be more frequent than those altering the same switch plus any additional switch. A binary combinatorial code therefore predicts specific sequences of transdetermination as a result of one-step transitions in that code. It also predicts a global orientation of transitions toward the most stable state of each binary switch (35,36). The particular binary code (Fig. 8) generated by the compartmental lines we predict on the blastoderm yields a large number of specific predictions, most of which are confirmed (Table 1).

Table 1. Predicted relative transdetermination frequencies derived from the chemical wave model applied to the blastoderm. $L_{1,2} \rightarrow A > L_{1,2} \rightarrow G$ means the model predicts transdetermination from the first or second leg to antenna is greater than to genital. Abbreviations are explained in the legend of Fig. 8.

Prediction	Status	Prediction	Status	Prediction	Status
$H \rightarrow W > H \rightarrow A$	T	$A \rightarrow W > A \rightarrow H$	T	$L \rightarrow W > L \rightarrow E$	T
$H \rightarrow W > H \rightarrow L_{1,2}$	T	$A \rightarrow L > A \rightarrow W$	F	$L_{1,2} \rightarrow W > L_{1,2} \rightarrow H$	T
$H \rightarrow W > H \rightarrow E$	T	$A \rightarrow Pb > G \rightarrow Pb$	?	$L > A > L \rightarrow E$	T
$H \rightarrow W > H \rightarrow Pb$	T	$A \rightarrow E > A \rightarrow W$	F	$L_{1,2} \rightarrow A > L_{1,2} \rightarrow G$	T
$W \rightarrow A > H \rightarrow A$	T	$A \rightarrow G > L_{1,2} \rightarrow G$	T	$L_2 \rightarrow G > L_3 \rightarrow A$	?
$W \rightarrow E > H \rightarrow E$	T	$A \rightarrow E > E \rightarrow A$	T	$L_1 \rightarrow Pb > L_1 \rightarrow G$	?
$W \rightarrow L_{1,2} > H \rightarrow L_{1,2}$	T	$A \rightarrow L_2 > L_2 \rightarrow A$	?T	$G \rightarrow A > G \rightarrow Pb$	T
$W \rightarrow L > W \rightarrow A$	T	$E \rightarrow W > E \rightarrow H$	T	$G \rightarrow A > G \rightarrow W$	T
$W \rightarrow L > W \rightarrow G$	T	$E \rightarrow A > E \rightarrow G$	T	$G \rightarrow L_{2,3} > G \rightarrow W$	?T
$W \rightarrow A > W \rightarrow G$	T	$E \rightarrow A > E \rightarrow L$	T	$G \rightarrow A > G \rightarrow L_{1,2}$	?T
$W \rightarrow E > W \rightarrow Pb$	T	$E \rightarrow W > E \rightarrow L$	T	$G \rightarrow A > A \rightarrow G$	T
$W \rightarrow E > W \rightarrow G$	T			$G \rightarrow L > L \rightarrow G$	T
$W \rightarrow E > W \rightarrow A$	?			$G \rightarrow H > G \rightarrow W$	?F

For example, transdetermination from haltere to wing should be more frequent than from haltere to antenna, since in both cases the first switch changes from 0 to 1, while in the conversion of haltere to antenna the second and fourth

switch must also change from 1 to 0. Of 30 predicted relative frequencies
which are unambiguously true or false, only two are false. Four of the 30
predictions reflect a subsidiary assumption that state 1 is more stable than
state 0 of each switch, hence that the transition of 0 to 1 is more frequent
in each switch than the transition of 1 to 0.

Hoemotic mutants convert one tissue to another in highly specific ways (10,
39). If each disc and disc compartment has made a unique combination of binary
choices recorded by means of switches, or "selector genes," then some homeotic
mutants might convert one compartment to another by altering the state of a
single switch. Garcia-Bellido (1) and Morata and Lawrence (17,40) have already
carried out an analysis of the mutants <u>engrailed</u>, <u>bithorax</u>, and <u>post-bithorax</u>
on the basis of this idea.

There are at least three general properties predicted of homeosis by a
binary code:

1) In most cases, any given "selector gene" is in the same state in
several discs. If a homeotic mutant converts the switch from one to another
state in one compartment, it might also do so in other compartments. A
<u>combinatorial</u> code predicts coordinated homeotic transformations (35,36).

2) The developmental programs in discs which are geographically distant
on the blastoderm can differ in the state of only a single switch (Fig. 8).
Thus, some homeotic mutants should transform between tissues widely separated
on the blastoderm.

3) If transdetermination in a disc results from altering any one of several
switches, but a homeotic mutant alters the state of a specific switch, then the
set of tissues to which one disc can transdetermine should be broader than, but
should include, the specific tissue to which one disc is transformed by a homeotic
mutant.

Homeotic mutants do exhibit these three general properties (39). Table 2
lists the homeotic mutants that cause transformations between tissues given
distinct combinatorial code names in Fig. 8. Many of the mutants involve
coordinated transformations of two distinct tissues. In addition, many involve
transformations between noncontiguous domains on the blastoderm fate map.
Comparison of Tables 1 and 2 confirms that the spectrum of transdeterminations

from each disc is broader than, but inclusive of, the effects of a specific homeotic mutant.

If a homeotic mutant acts by altering the states of a binary "selector gene," then an observed homeotic transformation of one tissue to another tissue might require a maximum of four switch alterations in a four digit binary code or a minimum of a single switch alteration. Similarly, if a homeotic causes coordinated transformations of two tissues to two other tissues (parallel transformation), both might, at a minimum, be due to the simultaneous altera-tion of the same single switch in the two tissues. However, homeotic trans-formation of one tissue into two different tissues (divergent transformation), or of two tissues into one tissue (convergent transformations) must require alteration of at least two switches. Table 2 shows that, in our specific binary code in Fig. 8, all but one single and coordinated parallel transformations require only a single switch alteration. All but one of the remaining trans-formations require the minimum two switch alterations.

The code in Fig. 8 derives from our predicted compartmental boundaries on the blastoderm. If compartmentalization truly reflects binary choices, then a more accurate code should result from an analysis of the actual detailed sequence and geometries of compartment formation in the early embryo. Correla-tion of such results with the body of data concerning sequential commitment in insect eggs provided by egg ligature, induction of double abdomen phenotypes, and other experiments (20,48,50), should not only test our model (51), but provide a more complete understanding of early developmental processes.

CONCLUSION

During development of <u>Drosophila</u> <u>melanogaster</u>, sequential commitment to alternative developmental programs occurs in neighboring groups of cells, and probably is reflected by formation of compartmental boundaries which progressively subdivide the early embryo and later the imaginal discs. The reasonable success of our chemical pattern model in predicting the locations and temporal order of compartmental boundaries, may lead to other projects showing that a uniform mechanism may act throughout development to determine the locations of successive developmental commitments. The success of the binary combinatorial code

Table 2. Observed homeotic transformation, and the code changes required for the code scheme in Fig. 8. A set of homeotic mutants causing the same transformation is represented by one member (10, 39): (1) Antennapedia, Antennapedix, aristatarsia; (2) Opthalmoptera, opthalmoptera (44), eyes-reduced: (3) tetraltera, Metaplasia, Haltere mimic; (4) extrasex combs, Extrasexcomb, reduplicated sex comb, sparse arista. Transformations with 1* and 2* require one additional switch to account for other transformations of that homeotic.

Mutant	Symbol	Transformation	Coordination	Code Change	Switches Required	Ref
Antennapedia(1)	Antp	antenna → leg 2	–	1010→1110	1	10, 39
Pointed wing	Pw	antenna → wing	–	1010→1111	2	39
Nasobemia	Ns	antenna → leg 2 / wing	parallel	1010→1110 / 1011→1111	1	44
dachsous	ds	eye → arista	–	1110→1010	1	45
Opthalmoptera(2)	OptG	eye → wing	–	1110→1111	1	39
Hexaptera	Hx	prothorax → mesothorax	–	1101→1111	1	39
podoptera	pod	wing → leg	–	1111→1110	1	39
tetraltera(3)	tet	wing → haltere	–	1111→0111	1	39
Contrabithorax	Cbx	wing → haltere / leg 2 → leg 3	parallel	1111→0111 / 1110→0110	1	39
Ultrabithorax	Ubx	haltere → wing / leg 3 → leg 2	parallel	0111→1111 / 0110→1110	1	43
tumorous head	tuh1,3	eye → genital / antenna → genital	parallel / divergent	1011→0011 / 1010→0010	1*	47, 39
lethal(3)III-10	l(3)III-10	haltere → wing / antenna → leg	parallel / divergent	0111→1111 / 0010→1010	1	34
lethal(3)XVI-18	l(3)XVI-18	genital → wing / genital → leg	parallel / divergent	0010→0110 / 1010→1110	1*	39
lethal(3)703	l(3)703	antenna → leg / genital → leg	parallel / divergent	1010→1110 / 0010→0110	1	34
lethal(3)1803R	l(3)1803R	genital → antenna / haltere → wing	divergent / parallel	0010→1010 / 0111→1111	1*	34, 39
proboscipedia	pb	proboscis → antenna / proboscis → leg	divergent	1000→1010 / 1000→1100	1	10, 39
extrasexcombs(4)	ecs	leg 2 → leg 1 / leg 3 → leg 1	convergent	1110→1100 / 0110→1100	2	39
Polycomb	Pc	antenna → leg	–	1010→1110	1*	39
lethal(4)29	l(4)29	leg 2 → leg 1 / leg 3 → leg 1	convergent	1110→1100 / 0110→1100	2*	10, 39

generated by the chemical patterns in accounting for transdetermination and homeotic mutants not only underscores the possibility that the logic of developmental commitments in <u>Drosophila</u> is written in a binary code, but also it yields a new view of sequential commitments in early embryogenesis which is open to experimental tests.

APPENDIX

This section deals with the analysis of the reaction-diffusion system.

In the chemical system (Eqs. 1), spatial patterns can spontaneously arise from an initial spatially homogeneous concentration profile by the selection from noise and amplification of perturbations with wavelengths in the neighborhood of some preferred wavelength. To determine the conditions under which this occurs, the behavior of the system in the vicinity of the spatially homogeneous steady state, $X = X_0$, $Y = Y_0$, where $F(X_0,Y_0) = G(X_0,Y_0) = 0$, is analyzed using a standard linearization procedure.

The system (Eqs. 1) is linearized about the spatially homogeneous steady state by substituting $X(r,t) = X_0 + x(r,t)$, $Y(r,t) = Y_0 + y(r,t)$, and retaining only terms up to first order in x and y in a Taylor expansion of $F(X,Y)$ and $G(X,Y)$. The resulting linear equations in x and y are

$$\partial x/\partial t = K_{11}x + K_{12}y + D_1\nabla^2 x$$
$$\partial y/\partial t = K_{21}x + K_{22}y + D_2\nabla^2 y$$

$$(A1)$$

These equations are solved by separating out the time dependence through the substitutions $x(r,t) = x'(r)e^{\lambda t}$, $y(r,t) = y'(r)e^{\lambda t}$, and diagonalizing the resulting pair of spatially dependent coupled equations. These two separated equations are Helmholtz-type equations whose solutions can be straightforwardly obtained in different coordinate systems (23,25). The complete space-time dependent solutions are sums of spatial modes or patterns, each with a characteristic temporal behavior. For example, the complete solution on a circle can be written

$$\begin{bmatrix} x(r,\phi,t) \\ y(r,\phi,t) \end{bmatrix} = \sum_{i = +,-} \sum_{j = 1}^{\infty} \sum_{n = 0}^{\infty} \begin{bmatrix} a_{nji} \\ b_{nji} \end{bmatrix} X$$

$$\exp \lambda_{nji} \, J_n(k_{nj}r)\cos n\phi \qquad (A2)$$

where J_n is the nth Bessel function and r and ϕ are the radial and angular coordinates, respectively, a_{nji} and b_{nji} are arbitrary constants to be determined by initial conditions. Each mode has two temporal eigenvalues, indicated by the sum $i = +,-$. The sum over j depends on the boundary conditions chosen.

For no-flux boundary conditions, the spatial gradient at the boundary must have zero component normal to the boundary (23). In a circle of radius r_0, this means that $\partial x(r,o,t)/\partial r = \partial y(r,o,t)/\partial r = 0$ at $r = r_0$. The zeros in the derivatives of $J_n(z)$ occur at particular values of the argument $z = z_{nj}$ (26). Therefore, the spatial mode $J_n(k_{nj}r)\cos n\phi$, which we abbreviate by J_{nj}, is obtained when the jth zero in the derivative of J_n occurs at the boundary; that is, when $k_{nj}r_0 = z_{nj}$. This fixes the value of k_{nj} associated with the mode J_{nj} for any given radius r_0. As the radius changes, the value of k_{nj} changes in inverse proportion.

The temporal behavior of the mode J_{nj} is determined by the dynamics through the dispersion relation between the temporal eigenvalues λ_{nji} and the spatial eigenvalue k_{nj}. For Eqs. A1 this relation has the form

$$\lambda_{nj\pm} = \frac{1}{2}\left[K_{11} + K_{22} - k_{nj}^2(D_1 + D_2) \right.$$
$$\pm \left\{ [K_{11} - K_{22} - k_{nj}^2(D_1 - D_2)]^2 \right.$$
$$\left. \left. + 4K_{12}K_{21} \right\}^{1/2} \right] \tag{A3}$$

As seen in Eq. A2, each mode J_{nj} behaves in time according to a sum of terms of the form $A \exp \lambda_{nj+}t + B \exp \lambda_{nj-}t$, where A and B are specified by initial conditions. Therefore, any mode with the real parts of λ_{nj+} and λ_{nj-} both negative will decay and disappear; a mode with either or both real parts positive will grow and create a spatial pattern.

On an ellipse, the solutions of Eq. A1 are

$$\begin{bmatrix} x(\xi,\eta,t) \\ y(\xi,\eta,t) \end{bmatrix} = \sum_{i=+,-} \sum_{j=1}^{\infty} \sum_{n=0}^{\infty} \begin{bmatrix} a_{nji} \\ b_{nji} \end{bmatrix} \times$$
$$\exp \lambda_{nji}t[Ce_n(\xi,s_{nj})ce_n(\eta,s_{nj}) +$$
$$S_{nji}Se_n(\xi,s_{nj})se_n(\eta,s_{nj})] \tag{A4}$$

Here ce_n and se_n are periodic cosine- and sine-elliptic Mathieu functions, respectively, of integral order and Ce_n and Se_n are the corresponding nonperiodic (or modified) Mathieu functions (25). ξ and η are the elliptical coordinates tracing out confocal ellipses and hyperbolae, respectively (Fig. 4), and $s_{nj} = h^2k_{nj}^2$, where h is one-half the interfocal distance of the ellipse. The constants a_{nji}, b_{nji}, and S_{nji} are determined by initial conditions. We use the abbreviations Ce_{nj} or Se_{nj} for the patterns $Ce_n(\xi,s_{nj})ce_n(\eta,s_{nj})$ or $Se_n(\xi,s_{nj})se_n(\eta,s_{nj})$, respectively.

At the boundary of the ellipse, $\xi = \xi_0$, the no-flux condition becomes $\partial Ce_n(\xi,s_{nj})/\partial\xi = 0$ or $\partial Se_n(\xi,s_{nj})/\partial\xi = 0$, analogous to the circular case. This condition fixes the value of the scaling factor k_{nj} for the pattern Ce_{nj} or Se_{nj}, and the temporal behavior of this pattern is determined by exactly the same dispersion relation (Eq. A3) as for the circular case. Those modes which grow in time will form spatial patterns on the ellipse; those which decay in time will not be seen.

If the following five conditions hold

$$\text{(i)} \quad K_{11} + K_{22} < 0$$

$$\text{(ii)} \quad K_{11}K_{22} - K_{12}K_{21} > 0$$

$$\text{(iii)} \quad (K_{11} - K_{22})^2 > -4K_{12}K_{21}$$

$$\text{(iv)} \quad D_1K_{22} + D_2K_{11} > 0$$

$$\text{(v)} \quad \left(\sqrt{\frac{D_1}{D_2}}\, K_{22} - \sqrt{\frac{D_2}{D_1}}\, K_{11} \right)^2 > -4K_{12}K_{21}$$

$$\text{(A5)}$$

spatial patterns will arise spontaneously. In such a system, one reactant diffuses more readily than the other, one catalyzes its own production while the other inhibits its own production, and one catalyzes the production of the second while the second inhibits the production of the first. Under these conditions, the temporal eigenvalues λ_{nj+} and λ_{nj-} will both be real and the larger of them, λ_{nj+}, will be positive only in the neighborhood of a particular value of k_{nj}, equal to $2\pi/l_0$ is the natural chemical wavelength of the system (Fig. 9). Since only those modes with postive Ce_{nj} or Se_{nj} will grow which satisfies the no-flux boundary condition with $k_{nj} \approx 2\pi/l_0$. Furthermore, since λ_{nj+} is real, this pattern will grow without oscillation.

For an ellipse with a given eccentricity $\varepsilon = 1/\cosh \xi_0$, the wave number k_{nj} for the pattern Ce_{nj} or Se_{nj} is inversely proportional to the interfocal distance, 2h, and therefore to the size of the ellipse (25), in direct analogy with the circular case. Therefore as the size increases, each k_{nj} will be scaled downward along the abscissa in Fig. 9, and different modes will appear in sequence as their respective k_{nj}'s enter the region of positive λ_{nj+}. However, since more and more modes will be compressed into the region of small

k_{nj}, eventually more than one mode will fall in the positive λ_{nj+} region. In this case, a superposition of modes might appear or, in a fully nonlinear system, a previously established mode might suppress a later mode even though both are allowed in the linear theory.

The particular pattern selected depends not only on the size of the ellipse, but also on its eccentricity or shape. In Fig. 10 we show a two-parameter space of all possible ellipses. The abscissa ξ specifies the eccentricity of the ellipse, $\cosh \xi = 1/\varepsilon$, and the ordinate specifies the parameter $s = h^2 k^2$, where 2h is the interfocal distance and the subscripts have been dropped. Since all allowed patterns have the same value of k $(=2\pi/1_0)$ s depends only on h^2, and therefore is a direct measure of the size of the ellipse. The lines labeled Ce_{nj} or Se_{nj} are the loci along which the indicated mode satisfies the no-flux boundary condition with $k_{nj} = 2\pi/1_0$ (28). The dotted portions of the lines are our extrapolations of the available data. Changes in size and shape of a smoothly growing ellipse can be plotted as a continuous trajectory in Fig. 10. Whenever such a trajectory intersects one of the mode lines, that mode will be selected and amplified. Therefore, the sequence of patterns that arise is determined by the sequence in which the mode lines are crossed by the growth trajectory. Note that since the region of amplification in Fig. 9 has a finite size, a given mode can be amplified in a small region on either side of its indicated mode line. If two mode lines are near one another both may be allowed in the linear theory, and superposition of the two or suppression of one by the other could occur.

The curve labeled W in Fig. 10 shows an estimate of the size and shape changes which occur in the growing wing disc, approximating it as a perfect ellipse (24). The predicted sequence of modes is Ce_{11}, Se_{11}, Ce_{21}, Se_{21}, Ce_{31}, and Ce_{01}. The final subscripts are dropped in the text.

In the linear approximation given by Eqs. A1, modes selected for amplification grow without bound. The nonlinear reaction-diffusion system

$$\frac{\partial X}{\partial t} = -AX + \frac{BY^n}{1 + Y^n} + D_1 \nabla^2 X$$

$$\frac{\partial Y}{\partial t} = -CX + \frac{D(Y^n + b)}{1 + Y^n} + D_2 \nabla^2 Y \qquad (A6)$$

is an example of a system which can create spatial patterns as described above, but in which these patterns grow only to a finite size. In this system, X inhibits both its own and Y's production, and Y catalyzes the production of both. Also, X diffuses more readily than Y.

In Eqs. A6, with $A = 7.8$, $B = 15.6$, $C = 1$, $D = 1.7$, $b = 0.2$, $n = 6$, $D_1 = 16$, $D_2 = 1$, this sytem has one spatially homogeneous steady state at $X_0 = Y_0 = 1$. Linearizing about this steady state and substituting the resulting linearization constants K_{ij} into the dispersion relation, Eq. A3, we find that λ_{nj+} is positive for k_{nj} between 0.7 and 1.0, corresponding to wavelengths between $l_{min} = 6.1$ and $l_{max} = 9.1$. In a one-dimensional domain $0 < r < L$, the linearization of Eqs. A6 would therefore amplify the pattern $\cos n\pi r/L$ henceforth called the "n-mode," in the range of lengths $\tfrac{1}{2}nl_{min} < L < \tfrac{1}{2}nl_{max}$. Computer simulations showed that the analogous nonlinear patterns appeared in the identical length ranges, although their shapes were slight distortions of pure cosines.

In symmetrical bilobed domains, represented by two identical rectangles joined by a short narrow isthmus, the even lengthwise modes, with antinodes at the join, appeared at the same overall lengths as for the one-dimensional case.

The range of lengths supporting the one-mode was compressed and shifted downward toward zero as the join was made narrower. The range of each higher odd mode shifted downward, overlapped that of the even mode below, and finally became coincident with the even mode's range in the limit of a very narrow join. These conditions tend to suppress the antisymmetrical modes if tissue growth is rapid enough, since the first anti-symmetrical mode could well be skipped entirely, and each symmetrical mode will become established and may not fully decay before growth has taken the system beyond the range of the overlapping anti-symmetrical mode.

For nonsymmetric bilobed shapes (with unequal rectangles) simulations showed that all lengthwise modes tended to appear at somewhat shorter overall lengths than their one-dimensional counterparts. However, the exact sequence of modes depends critically on the geometry, making predictions difficult.

REFERENCES AND NOTES

1. Garcia-Bellido, A., Ciba Found. Symp. 29 (1975) 161.

2. Garcia-Bellido, A., Ripoll, P., and Morata G., Dev. Biol. 48 (1976) 132.

3. Baker, W., in preparation.

4. Steiner, E., Arch. Entwickl. Org. 180 (1976) 9.

5. Wieschaus, E., and Gehring, W., Dev. Biol. 50 (1975) 249.

6. Dubendorfer, K, thesis, University of Zurich (1977).

7. Crick, F. H. C., and Lawrence, P. A., Science 189 (1975) 340.

8. Turner, F. R., and Mahowald, A. P., Dev. Biol. 50 (1976) 95.

9. Chan, L. J., and Gehring, W., Proc. Natl. Acad. Sci. U.S.A. 68 (1971) 2217;
    Illmensee, K., Arch. Entwickl. Org. 170 (1972) 267.

10. Nothiger, R., and Gehring, W. in Developmental Systems II: Insects, S. Counce
    and C. H. Waddington, Eds. (Academic Press, New York, 1973), p. 161.

11. Garcia-Bellido, A., and Merriam, J. R., J. Exp. Zool. 170 (1969) 1.

12. Lawrence, P., and Morata, G., Dev. Biol. 56 (1977) 40.

13. Morata, G., and Ripoll, P., ibid. 42 (1975) 221.

14. Bryant, P., Ciba Found. Symp. 29 (1975) 71; Bryant, P., J. Exp. Zool. 193
    (1975) 49.

15. Haynie, J., and Bryant, P., Nature (London) 259 (1976) 659.

16. Morata, G., J. Embryol. Exp. Morphol. 34 (1975) 19; Garcia-Bellido, A. and
    Santamaria, P., Genetics 72 (1972) 87.

17. Morata, G., and Lawrence, P. A., Nature (London) 255 (1975) 614; Morata, G.,
    and Lawrence, P. A., ibid. 265 (1977) 211.

18. Turing, A. M., Philos. Trans. R. Soc. London, Ser. B 237 (1952) 37.

19. Gmitro, J. I., and Scivin, L. E. in Intracellular Transport, K. B. Warren,
    Ed. (Academic Press, New York, 1966).

20. Meinhardt, H., J. Cell Sci. 23 (1977) 117.

21. Nicolis, G., and Prigogine, I., Self-Organization in Nonequilibrium Systems
    (Interscience, New York, 1977).

22. For reaction times τ of the order of 100 seconds and diffusion constants
    of the order of 10^{-6} cm^2/sec, the characteristic "diffusion length" of the
    system is $(D\tau)^{\frac{1}{2}} = 100$ μm.

23. Morse, P. M., and Feshback, H., Methods of Theoretical Physics (McGraw-Hill,
    New York, 1953).

24. Auerbach, C., <u>Trans. R. Soc.</u> <u>58</u> (1933) 787; Madhavan, M., personal communication.

25. McLachlan, N. W., <u>Theory and Applications of Mathieu Functions</u> (Clarendon, Oxford, 1947).

26. Abramowitz, M., and Stegun, I. A., <u>Handbook of Mathematical Functions</u> (Government Printing Office, Washington, D.C., 1972).

27. Jahnke, E., Emde, F., and Lusch, F., <u>Table of Higher Mathematical Functions</u> (McGraw-Hill, New York, 1960).

28. King, M. J., and Wiltse, J. C., <u>Derivative Zeros and Other Data Pertaining to Mathieu Functions</u> (Johns Hopkins Radiation Laboratory Technical Report No. AF57, Baltimore, Md., 1958).

29. Morata, G., personal communication.

30. Demerec, M., <u>The Biology of Drosophila</u> (Wiley, New York, 1950).

31. Stimulating a fully nonlinear model with a natural wavelength in a growing one-dimensional domain, Harrison and Lacalli (personal communication) found that after t   first half-wavelength asymmetrical mode, several symmetrical modes appeared. We assume that on the egg, patterns analogous to Ce_1, Ce_2, Ce_4 and Ce_6 form in succession.

32. Babloyantz, A., and Hierneux, J., <u>J. Math. Biol.</u> <u>37</u> (1975) 637.

33. Herschkowitz-Kaufman, M., <u>Bull. Math. Biol.</u> <u>37</u> (1975) 589.

34. Shearn, A., Rice, T., Garen, A., and Gehring, W., <u>Proc. Natl. Acad. Sci. U.S.A.</u> <u>68</u> (1971) 2594.

35. Kauffman, S. A., <u>Science</u> <u>181</u> (1973) 310.

36. Kauffman, S. A., <u>Ciba Found. Symp.</u> <u>29</u> (1975) 201.

37. Kauffman, S. A., <u>Am. Zool.</u> <u>17</u> (1977) 631.

38. Hadorn, E. in <u>Major Problems in Developmental Biology</u>, M. Locke, Ed. (Academic Press, New York, 1966), p. 85.

39. Ouweneel, W. J., <u>Adv. Genet.</u> <u>18</u> (1976) 179.

40. Lawrence, P., and Morata, G., <u>Dev. Biol.</u> <u>50</u> (1976) 321.

41. Ouweneel, W. J., <u>Acta Embryol. Exp.</u> <u>95</u> (1970) 119.

42. Lewis, E. B., <u>Drosophila Inform. Serv.</u> <u>30</u> (1956) 130; Roberts, P., <u>Genetics</u> <u>49</u> (1964) 593.

43. Lewis, E. B. in <u>The Role of the Chromosomes in Development</u>, M. Locke, Ed. (Academic Press, New York, 1964).

44. Gehring, W. J., Arch. Julius Klaus-Stift. Verebungsforsch. Sozialanthropol. Rassenhyg. 41 (1966) 44.

45. Stepshin, V. P., Ginter, E. K., Genetika 8, (8) (1972) 93.

46. Bull, A. L., J. Exp. Zool. 161 (1966) 221.

47. Postlethwait, J. H., Bryant, P., and Schubiger, G., Dev. Biol. 29 (1972) 337.

48. Schubiger, G., Moseley, R. C., and Wood, W. J., Proc. Natl. Acad. Sci. U.S.A. 74 (1977) 2050.

49. Hotta, Y. and Benzer, S., ibid. 73 (1976) 4154.

50. Sander, K., Adv. Insect Physiol. 12 (1976) 125.

51. Ligation of Drosophila and other insect eggs at progressively later cleavage stages causes a progressively narrower gap of missing larval or adult structures centered about the line of ligation (48,50). The bicaudal mutant of Drosophila produces mirror image abdomen with all or only the posterior few segments (46,50). Similar double abdomen in Smittia can be produced by puncture or irradiation of the anterior egg pole (50). Elegant single and double gradient models (20,50) have been proposed to account for these phenomena. Although our model was not constructed with these data in mind, it can predict at least their major qualitative features. We shall discuss this in detail elsewhere.

52. Supported in part by grants from the National Institutes of Health (1-R01-GM-22341-01 and the National Science Foundation (BMS-75-11917). Part of this work was done while S. K. and K. T. were at the Laboratory of Theoretical Biology, National Cancer Institute.

Stuart A. Kauffman, Ronald M. Shymko,
Department of Biochemistry and Biophysics
School of Medicine
University of Pennsylvania
Philadelphia, PA 19104

and

Kenneth Trabert
Environmental Protection Agency
Washington, D.C.